ENGLISH
CATHEDRAL
AND
MONASTIC
CARPENTRY

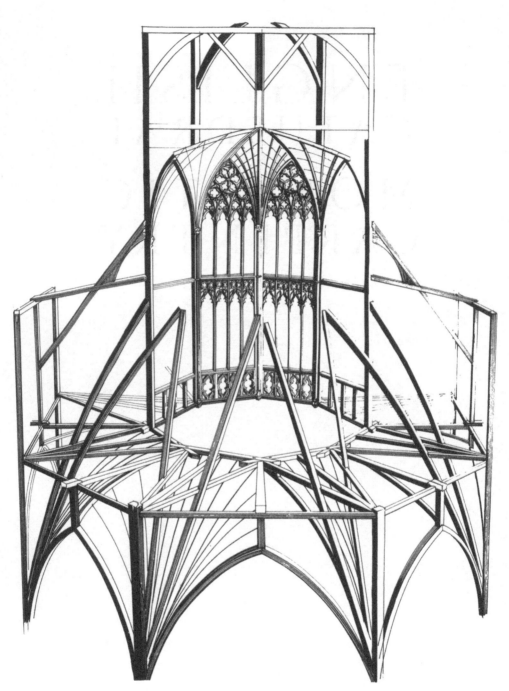

Ely Cathedral, the framing of the lantern.

ENGLISH
CATHEDRAL
AND
MONASTIC
CARPENTRY

CECIL A. HEWETT

Front cover image: Peterborough Cathedral. (Colin Underhill/Alamy Stock Photo)

First published in 1999 by Sutton Publishing
This paperback edition published in 2023

The History Press
97 St George's Place, Cheltenham,
Gloucestershire, GL50 3QB
www.thehistorypress.co.uk

British Library Cataloguing in Publication Data.
A catalogue record for this book is available from the British Library.

ISBN 978 1 80399 477 2

Typesetting and origination by The History Press
Printed and bound in Great Britain by TJ Books Limited, Padstow, Cornwall.

Trees for Life

CONTENTS

LIST OF PLATES

1. The spire of Salisbury Cathedral
2. From Rayleigh Castle, Essex, a post excavated in 1910
3. A notched lap joint of archaic profile from the belfry of Navestock church, Essex
4. A secret notched lap joint without refined profile from Romsey Abbey, Hampshire
5. Portrait of Adam Lock, master mason, from Wells Cathedral
6. An open notched lap joint of refined profile, from Wells Cathedral
7. Woodcut, c.1515, showing two types of axes in use at that time
8. Mortise joint from the Barley Barn, Cressing Temple, Essex
9. An oblique mortise from the Barley Barn, Cressing Temple, Essex
10. The timber roof over the Chapter House, Wells Cathedral

LIST OF FIGURES

BIBLIOGRAPHY

Aylmer and Cant 1977: G. Aylmer and R. Cant, *A History of York Minster* (O.U.P., 1977).
Backinsell: W. G. C. Backinsell, 'The Medieval Clock in Salisbury Cathedral' (Historical Monograph 2, *South Wiltshire Industrial Archaeology Society*).
Banister Fletcher 1896: Sir F. Banister Fletcher, *A History of Architecture* (Batsford, 1896).
Berger and Libby 1967: R. Berger and W. F. Libby, 'European Medieval Architecture Series' (*Radiocarbon 9*, Yale 1967, pp. 486-90).
Bilson 1928: J. Bilson, 'Notes on the Earlier Architectural History of Wells Cathedral' (*Archaeological Journal* LXXXV, 1928, pp. 23-68).
Bony 1979: J. Bony, *The English Decorated Style* (Phaidon, 1979).
Bowie 1959: T. Bowie (ed.), *The Sketchbook of Villard de Honnecourt* (1959).
Britton 1821: J. Britton, *The History and Antiquities of the Cathedral Church of Oxford* (1821).
Chapman 1907: F. R. Chapman, *The Sacrist Rolls of Ely* (C.U.P., 1907).
Colchester and Harvey 1974: L. S. Colchester and J. H. Harvey, 'Wells Cathedral' (*Archaeological Journal* CXXXI, 1974, pp. 200-214).
Colvin 1948: H. M. Colvin, 'Abbey Dore' (*Transactions of the Woolhope Naturalists' Field Club* XXXII, 1948, pp. 235-7).
Colvin 1963: H. M. Colvin (ed.), *The History of the King's Works* (HMSO, 1963).
Colvin 1978: H. M. Colvin, *A Biographical Dictionary of British Architects, 1600-1840* (Murray, 1978).
Davis 1966: R. H. C. Davis, 'The Norman Conquest' (*History* LI, 1966, pp. 279-286).
Dale 1956: A. Dale, *James Wyatt* (Blackwell, 1956).
Dickinson 1978: J. C. Dickinson, *Cartmel Priory Church* (1978).
Downes 1959: J. K. Downes, *Hawksmoor* (Zwemmer, 1959).
Drinkwater 1964: N. Drinkwater, 'Old Deanery, Salisbury' (*Antiquaries Journal* XLIV, 1964, pp. 41-59).

Dunlop 1972: D. C. Dunlop, *Lincoln Cathedral* (Pitkin, 1972).

Everett 1944: C. R. Everett, 'Notes on the Decanal and other Houses in the Close of Sarum' (*Wiltshire Archaeological and Natural History Magazine*, 50, 1944, pp. 425-445).

Fletcher and Haslop 1971: J. M. Fletcher and F. W. O. Haslop, 'The West Range at Ely and Its Romanesque Roof' (*Archaeological Journal* CXXVI, 1971, pp. 171-6).

Fletcher and Hewett 1969: J. M. Fletcher and C. A. Hewett, 'Medieval Timberwork at Bisham Abbey' (*Medieval Archaeology* XIII, 1969, pp. 220-4).

Fletcher and Spokes 1964: J. M. Fletcher and P. S. Spokes, 'The Origin and Development of Crown-Post Roofs' (*Medieval Archaeology* VIII, 1964, pp. 153-83).

Focillon 1963: H. Focillon, *The Art of the West* (Phaidon, 1963).

Forrester 1972: H. Forrester, *Medieval Gothic Mouldings* (Phillimore, 1972).

Francis 1913: E. B. Francis, 'Rayleigh Castle, new facts in its history and recent excavations on site' (*Transactions of the Essex Archaeological Society* XII, n.s. 1913, pp. 147-61).

Gardner 1955: J. S. Gardner, 'Coggeshall Abbey and its early brickwork' (*Journal of the British Archaeological Association*, 3rd series, XVIII, 1955, pp. 19-32).

Geddes 1982: J. Geddes, 'The Construction of Medieval Doors, Woodworking Techniques before 1500' (Archaeological Series 7, *B.A.R. International Series* 129, National Maritime Museum, Greenwich, 1982).

Gibson 1976: A. V. B. Gibson, 'The Medieval Aisled Barn at Parkbury Farm, Radlett – Thirteenth century rafters re-used' (*Hertfordshire Archaeology* 4, 1974-6, pp. 158-163).

Harvey 1974: J. H. Harvey, *Cathedrals of England and Wales* (Batsford, 1974).

Hewett 1968: C. A. Hewett, 'Developments in Carpentry Illustrated by Essex Millwrighting' (*Art Bulletin* L, 1968).

Hewett 1969: C. A. Hewett, *The Development of Carpentry, 1200-1700, an Essex Study* (David and Charles, 1969).

Hewett 1974: C. A. Hewett, *English Cathedral Carpentry* (Wayland, 1974).

Hewett 1977: C. A. Hewett, 'Scarf Jointing during the later Thirteenth and Fourteenth Centuries, and a reappraisal of the Origin of Spurred Tenons' (*Archaeological Journal* 134, 1977, pp. 287-296).

Hewett 1980: C. A. Hewett, *English Historic Carpentry* (Phillimore, 1980).

Hewett 1982: C. A. Hewett, *Church Carpentry, A Study based on Essex Examples* (Phillimore, 1982).

Hewett and Smith 1972: C. A. Hewett and J. R. Smith, 'Faked Masonry of the Mid-13th Century in Navestock Church' (*Essex Journal*, Autumn 1972, pp. 82-5).

Hewett and Tatton-Brown 1976: C. A. Hewett and T. Tatton-Brown, 'New Structural Evidence regarding Bell Harry Tower and the South-east Spire at Canterbury' (*Archaeologia Cantiana* XCII, 1976, pp. 129-136).

Hope 1900: W. H. St John Hope, *The Architectural History of the Cathedral Church and Monastery of St Andrew at Rochester* (London, 1900).

Lang 1956: J. Lang, *Rebuilding St Paul's after the Great Fire of London* (O.U.P., 1956).

McDowall, Smith and Stell 1966: R. W. McDowall, J. T. Smith and C. F. Stell, 'Westminster Abbey, the Timber Roofs of the Collegiate Church of St Peter at Westminster' (*Archaeologia* C, 1966, pp. 155-74).

Macmahon 1970: K. A. Macmahon, *The Pictorial History of Beverley Minster* (Pitkin, 1970).

Marshall 1951: G. Marshall, *Hereford Cathedral, its Evolution and Growth* (Worcester Press, 1951).

Martin 1935: A. R. Martin, 'The Greyfriars of Lincoln' (*Archaeological Journal*, XCII, 1935, pp. 42-63).

Morgan 1967: F. C. Morgan, *A Short Account of the Church of Abbey Dore* (Leominster, 1967).

Parkin 1970: E. W. Parkin, 'Cogan House, St Peter's, Canterbury' (*Archaeologia Cantiana* LXXXV, 1970, pp. 123-138).

Pevsner 1965: Sir N. Pevsner, *The Buildings of England – Essex* (Penguin, 1965).

Pevsner 1970: Sir N. Pevsner, *The Buildings of England – Cambridgeshire* (Penguin, 1970).

Pierce 1965: J. A. Repton, *Norwich Cathedral at the end of the Eighteenth Century*, ed. by R. S. Pierce (Gregg Press, 1965).

Powell 1974: A. C. Powell, *The Hospital of the Blessed Mary*, Chichester (Regnum Press, Chichester, 1974).

Price 1774: R. Price, *A Description of that Admirable Structure, the Cathedral Church of Salisbury* (London, 1774).

Rackham, Blair and Munby 1978: O. Rackham, W. J. Blair and J. T. Munby, 'The Thirteenth Century Roofs and Floor of Blackfriars Priory at Gloucester' (*Medieval Archaeology* XXII, 1978, pp. 105-22).

Rigold 1977: S. E. Rigold, 'Romanesque Bases in and South-East of the Limestone Belt' (*Ancient Monuments and Their Interpretation*, essays presented to A. J. Taylor, ed. by Apted, Gilyard-Beer and Saunders, Phillimore, 1977).

Rowe 1876: R. R. Rowe, 'The Octagon and Lantern of Ely Cathedral' (*Trans. Royal Institute of British Architects*, 1875-6, pp. 69-83).

St Albans 1952: *St Albans Cathedral, Historical Monuments* (HMSO, 1952).

Salzman 1952: L. F. Salzman, *Building in England down to 1540, a documentary history* (O.U.P., 1952).

Scott 1863: G. G. Scott, *Gleanings from Westminster Abbey* (1863).

Singer 1956: C. Singer (ed.), *History of Technology* (Oxford, 1956).

Steward 1875: D. J. Steward, 'Notes on Norwich Cathedral' (*Archaeological Journal* XXXII, 1875).

Stranks 1972: C. J. Stranks, *Durham Cathedral* (Pitkin, 1972).

Venables 1883: The Rev. Precentor Venables, 'The Architectural History of Lincoln Cathedral' (*Archaeological Journal* XL, 1883, pp. 159-92 and 377-418).

V.C.H. Berks.: *The Victoria County History of Berkshire* (1923).

V.C.H. Essex: *The Victoria County History of Essex*.

Wallis 1970: Canon J. E. W. Wallis, *Lichfield Cathedral*, revised and enlarged by O. Hedley (Pitkin, 1970).

Whitham 1982: J. A. Whitham, *The Church of St Mary of Ottery* (Gloucester, 1982).

Wood 1965: M. Wood, *The English Mediaeval House* (Dent, 1965).

INTRODUCTION

This collection of drawings comprises examples of carpentry mainly in English Cathedrals, Abbeys, Priories, and their precincts, which have received less attention than they deserve. In many cases it is known, and in others it can reasonably be conjectured that these buildings are by masters who are well known from their cathedral and royal works; consequently they are valuable for dating mouldings and carpentry technology. The text is an outline only, having been prepared before the author became seriously ill. There are many more drawings which could have been used, but this would have made the book twice as long; therefore only the most important ones are dealt with. For example, Boxgrove Priory is included for the importance of the crown post. It cannot at present be amplified, but it is hoped that the drawings, some of which are of recent date, will be of value to the student. Some information about certain cathedrals, in particular Lincoln and Wells, which was not available at the time of the publication of *English Cathedral Carpentry*, is included also.

The schools of architecture are not yet trained in the skills of carpentry and its tools; the author feels that unless and until architects working with our most valuable buildings consult with historians, we shall be left with nothing for future generations. Already there has been grievous loss of historic structures in the unnecessary replacement of roof timbers in Westminster Abbey and the cathedrals of Norwich and

Worcester. It appears that money is available to ensure that this desecration of the ultimate art of carpentry can be continued.

There is very little time left, as from the conception of this book in 1970 to its present printing in 1985 we have lost much of value in three of our major national buildings. It is to be hoped that publication of the facts as to what may yet be saved may be useful in arousing sufficient interest in the subject to ensure the safety of the little that is left.

Mouldings are of particular interest to the author at present; a section of this book is devoted to the subject, and a detailed study is planned for the future.

RIDGED MAIN-SPAN ROOFS

Peterborough Cathedral Precinct, the 'Norman Hall' (*Fig. 1*)

This is probably the hall of the Norman infirmary built during the period of William Waterville, who was Abbot from 1155 to 1177. The apexes of the rafters are bridled and pegged, and the ends of both upper and lower collars are notch-lapped into the rafters. The tops of the ashlars are also notch-lapped, and the profiles of all the notched lap joints are of the 'archaic' type. The feet of the ashlars are fitted into housings that have sunken base abutments, and pegged – a device obviously calculated to ensure their stability in the vertical position necessary for the performance of their function. The feet of the rafters are tenoned into the outer ends of the sole-pieces. All the sole-pieces are trenched across their soffits in order that they may 'house' the wall-plate, which is of relatively small section and laid along the centre-line of the masonry. There were no tie-beams, as in Soignies Abbey, the chapel at Harlowbury, Essex (Hewett 1980, 47-8) and Chipping Ongar church, Essex (Hewett 1982, 3-4). It is remarkable that this roof has survived for so long. This must be largely due to the two tiebeams which have been inserted into it at some early and unknown date.

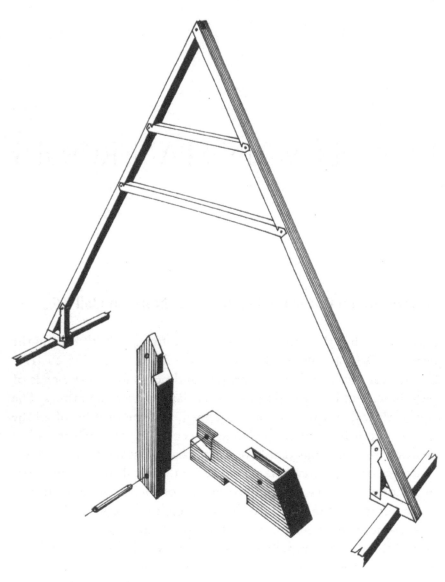

1. Peterborough Cathedral precinct, the roof of The Norman Hall.

Waltham Abbey, Essex, the Nave

The parish church of Waltham Abbey is the surviving part of Holy Cross, the nave once having formed a grand narthex to the Abbey church; this stage is not very clear. 'As for dates, the earliest grooved columns seem to be those at Durham of *c*.1095-1100. Those at Norwich are datable before 1119. The plain, heavy, ground-floor capitals at Waltham Abbey look more 11th than 12th century. However, the arches have zigzag decoration from the beginning, and zigzag does not occur anywhere in England before *c*.1105-10, so that date may mark the beginning of the western part, including their gallery. The clerestory was then erected on the north side, then that on the south, and finally the eastern bays were tackled and erected, including their clerestory and the arches to the crossing and transept. They may well belong to the mid-12th century, or even a little later' (Pevsner 1965, 403). Mr. S. E. Rigold dated it to the 1130s: 'Waltham, Essex, where the piers are integral with the suppressed apse, recently excavated ... consistent with the end of Henry I's reign' (Rigold 1977, 117).

It seems to have retained much of its original roof until 1807, when the component timbers were cut about sufficiently to produce a low-pitched, ridged and gabled roof – now slate clad – with king-post trusses and raking struts. Apparently, all of the original tie-beams, which were spaced at centres of one yard, were retained to support a boarded timber ceiling (which is floored within the roof-space, making examination difficult). The tie-beams were tenoned, without any refinements of jointing, into an eaves-plate which constituted the seatingtimber for the feet of the rafters – all of which were common – and the whole was single-framed. Measured drawings of the surviving timbers, some of which remain *in situ*, enable the reconstruction in Fig. 2 to be made with certainty. The wall-plates were partly embedded in the masonry string-course at mid-wall, and worked to a hewn face-fillet that registered each tie-beam. Ashlar-pieces were fitted in lap-joints designed sensitively to resist compression, and double collars were fitted by notched lap joints of archaic profile, to resist extension – as were the soulaces fitted beneath the lower collars.

Peterborough Cathedral, the North-West Portico

At Peterborough Cathedral seven rafter-couples survive, forming a roof above the north-west portico; their date is uncertain, but the

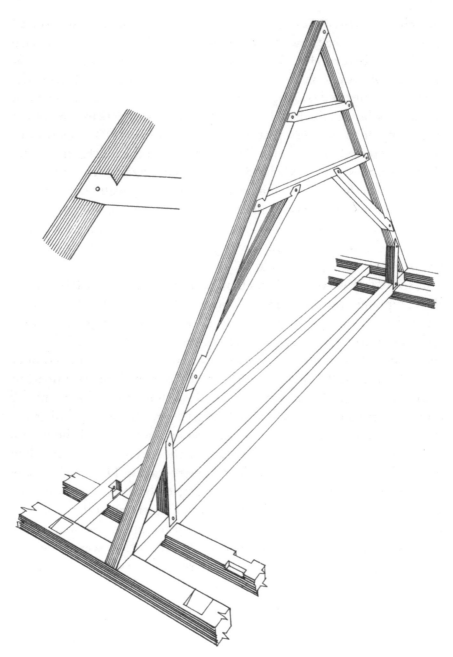

2. Waltham Abbey, a reconstruction of the nave roof.

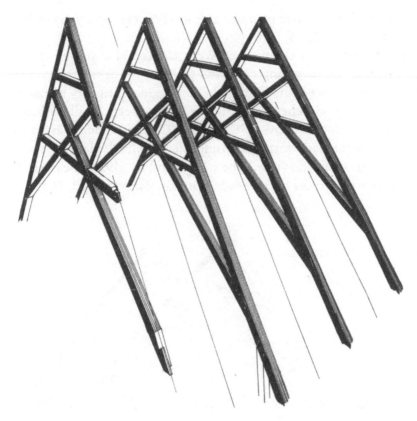

3. Peterborough Cathedral, the roof of the north-west portico.

entire nave was constructed between *c.*1155 and 1175 and the west front between 1193 and 1230 (Harvey 1974, 129). In view of the jointing used – open notched lap joints with archaic profiles – these couples relate, as R. Reuter first pointed out, to the nave and its construction period. The feet of the rafters were removed when they were re-used in their present position, as indicated in Fig. 3. All have double collars and scissor-braces (R. Reuter of Darmstadt Technische Hochschule, personal communication, 1964).

Wells Cathedral, the Nave – the first phase

'The Master of Wells, therefore, has to be envisaged as a young man already showing genius in or very soon after 1175, and capable of producing the complete plan and at least choir elevations, by or before 1180' (Harvey 1982, 55).

The high-roof of the nave of Wells Cathedral is among the most important of all surviving cathedral carpentry works. Its two phases of construction were divided by the 'break' in continuity of masonry which Mr. L. S. Colchester has shown to have been, in all probability, coincident with the Interdict, 1209-1213 (Colchester and Harvey 1974, 202-3). The same authorities consider that the likely date for the start of building operations would be *c.*1180, and by 1192 the eastern bays of the nave roof must have been in position (Harvey 1972, 163). Fig. 4

4. Wells Cathedral, the high-roof of the nave east of the 'break', showing the earlier form of notched lap joint used in this section only.

shows the main-span roof as built during the initial phase. The design was derived from that of the roof at Holy Cross, Waltham Abbey, with the important difference of mountants that were designed to resist extension being fitted between the two collars. Open notched lap joints were used at all points, including the ashlar-pieces, and these joints were given the 'refined' angle of entry that had been found less likely to tear or split under strain. Later the roofs at Wells were converted from eaves to parapets, and at that time the evidence for the original eaves-framing was lost. There were tie-beams at close intervals, and these survive, now embedded in the masonry of the parapets. The original work was executed under Master Adam Lock, medieval architect, who would have been responsible for both masonry and carpentry.

Lincoln Cathedral, St Hugh's Choir

'This roof seems, upon all grounds, both structural and documentary, to be the original one built under St Hugh between 1195 and his death in 1200' (Dr. J. H. Harvey, personal communication, 1 February 1981).

It has bay-lengths of two common-couples, defined by a total of 19 tie-beams, each of which originally had queen-posts chamfered into octagonal forms, as shown in Fig. 5. These were fitted by means of barefaced lap-dovetails at their feet — by virtue of which it has always been possible to renew them. This was a roof designed with double collars, above which were fitted chase-tenoned raking-struts; below the lower collars were fitted truncated under-rafters which were scissored by notch-lapped soulaces – their entry angles being of archaic profile. It appears that the basis of this design was the creation of a stable plane-figure beneath the collar and above the tie-beams, which was further stiffened by the inserted queen-posts, as had been used also at Salisbury Cathedral by *c*.1237, and later at Tewkesbury Abbey. The whole was tied within its own width by means of the truncated under-rafters, which were diagonally tied in a shallower pitch than that of the roof. The queen-posts form a remarkable feature in England, where they were not normally employed at this period, but in the European context of the Burgundian monk who was to become St Hugh, it is possible to indicate a similar although different system having been in use at the

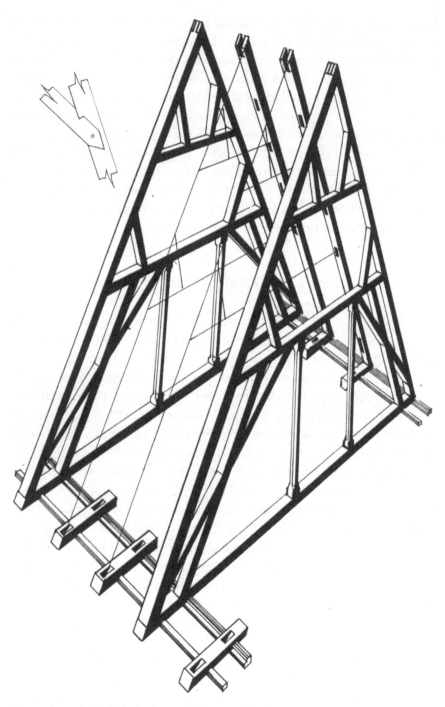

5. Lincoln Cathedral, the high-roof of St Hugh's Choir.

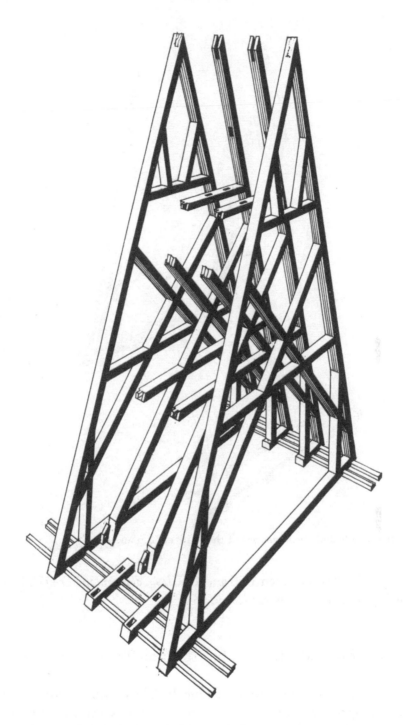

6. Lincoln Cathedral, the high-roof of the north choir transept.

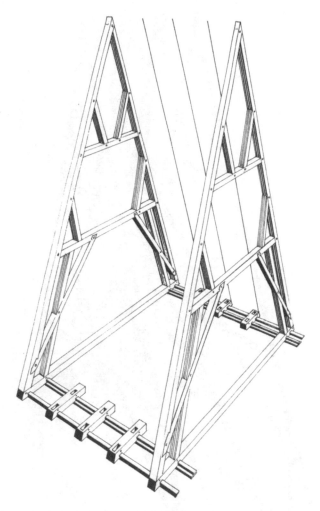

7. Lincoln Cathedral, the high-roof of the south choir transept.

Abbey of Noirlac between 1150 and 1160 (*No. d'Inventaire D 4553, Centre de la Recherche des Monuments Historiques*).

Lincoln Cathedral, the North Choir Transept (*Fig. 6*)

This is unique among the roofs of Lincoln. The measured span is 29 ft., and it comprises 30 rafter-couples with a further seven jack-rafters up the valleys. The bay lengths are, again, of two common-couples between tie-beams, the whole being mounted upon wall-plates that are double and have hewn face-fillets at the internal edges of the inner pair – clearly original timbers. The couples have scissor-braces that cross the lower collars and chase-tenon into the rafters above. At their lower ends they have notched lap joints of archaic profile, where they cross tall in-canted ashlar-pieces by halvings; the archaic entries are splitting at many points. The upper collars have raking-struts with chase-tenons at both ends; the couples were therefore designed to act in compression, supported at frequent intervals, based upon a rigidly inflexible eaves triangulation – be it tied or sole-plated. According to the account of one past Subdean, work was still in progress at this point when St Hugh died in 1200 (Cook 1970, 5).

Lincoln Cathedral, the South Choir Transept

'South Choir Transept – probably soon after St Hugh's death, therefore between 1200 and 1210' (Dr. J. H. Harvey, personal communication, 1 February 1981). This roof closely resembles that of St Hugh's Choir, and is illustrated in Fig. 7. It was designed in bays comprising three common-couples between tie-beams, and having truncated under-rafters, as there, that were scissored by notch-lapped soulaces having archaic profiles. Here too the wall-plating was doubled, a narrow fillet-like external timber with another having a hewn face-fillet, over which the sole-plates were registered. A total of 28 couples exist, with an additional seven jack-couples ascending the valley-rafters. No queen-posts were ever fitted in this range – possibly owing to its narrower span, a measured 29 ft. The wall-plates described are not entirely originals, but sufficient survives to establish their type beyond reasonable doubt.

Wells Cathedral, the Nave – the second phase

The roof frames used during the second phase of building after the Interdict of 1209-13 were identical with those of the first phase but for the adoption of *secret* notched laps at all points except the crown-pieces, where secret barefaced lap-dovetails were used. This phase is illustrated in Fig. 8. The date of completion was probably about 1230 (Harvey 1974, 241; Bilson 1928, 67-8).

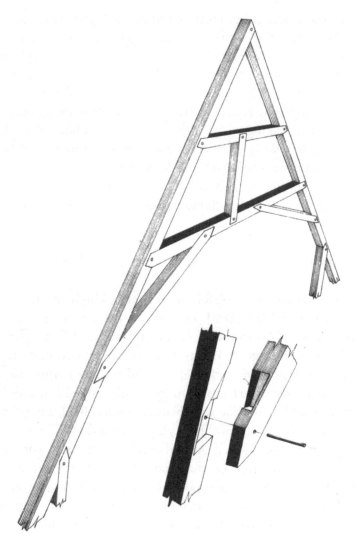

8. Wells Cathedral, the high-roof of the nave west of the 'break', showing the 'secret' form of notched lap joint not used in the earlier part.

Hereford Cathedral, the Lady Chapel

The roof shown in Fig. 9 is the one surviving over the Lady Chapel of Hereford Cathedral, which is dated to 1217-25 (Harvey 1974, 135). No records refer to the roof save those of the restoration undertaken

9. Hereford Cathedral, the roof of the Lady Chapel, with the present roof at right.

by James Wyatt between 1786 and 1796, when the raising of the pitch –
shown by the height of an added series of king-posts (at lower right of
the drawing) was probably effected. The old roof, which is of uncertain
date, was a development of the scissor-braced category of frames, in
which the principal-rafters were mounted on the feet of the scissors – a
technique later exploited by Sir Christopher Wren. Six side-purlins and
a diagonally-set ridge-piece were originally fitted, and the sole-pieces
were much elongated.

Coggeshall Abbey, Essex, the Chapel of St Nicholas

The abbey at Little Coggeshall still retains, intact, its *capella extra portas*, or chapel-without-the-gate. It has survived long periods of use as a farm building, and also a 19th-century restoration, without losing its Early English roof, comprising 25 rafter couples with collar and scissor braces – a pair of which is shown in Fig. 10, in which the joints spaced along each rafter are shown separated to the right. The recorded history of the abbey indicates that the chapel would have been built during the time of Abbot Benedict, 1218-23 (Gardner 1955, 20-21). The Royal Commission also ascribed it to *c.*1220.

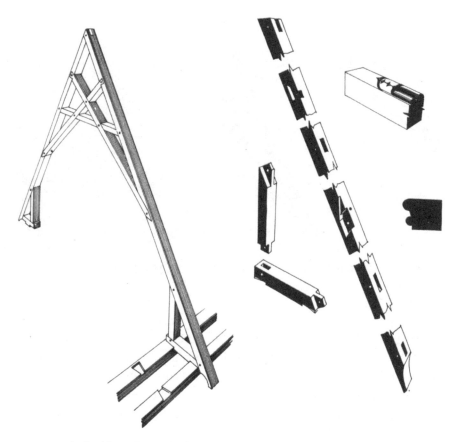

10. Coggeshall Abbey, the roof of the *capella extra portas*.

Chichester Cathedral, the Nave

The high-roof of Chichester Cathedral is continuous from east to west, unchanged by the structural intervention of the crossing tower, implying that it resulted from a single and smoothly-executed building operation. It must date from after the great fire of 20 October 1187, after which the church was re-roofed under Bishop Seffrid and the vaults added; reconsecration took place in 1199. One bay of this roof is drawn as Fig. 11, which is the bay terminating at the west gable. It is of a design that is unique to date in this field, with very tall crown-posts,

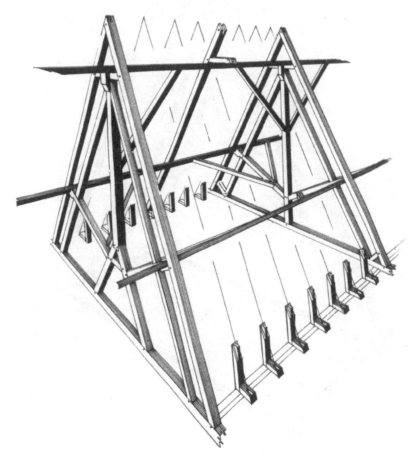

11. Chichester Cathedral, the high-roof of the nave.

common-collars and collar-purlin, raking-struts, side-purlins, and under-rafters terminating at the heads of the crown-posts. The bays comprise seven couples each, excluding the principals, and the wall-plates are single and laid along the internal masonry arrises. The dating of this work must depend upon that of the introduction of the jowl to the jointed ends of timbers, because pronounced jowls were used for both crown-posts and raking-struts. The 'upstand', the forerunner of the profiled jowl, was in use until *c*.1255 by the Knights Templar at

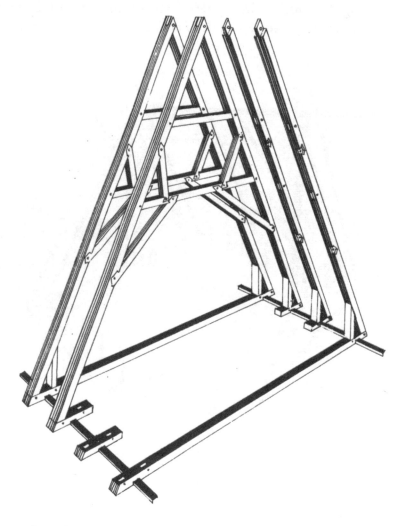

12. Lincoln Cathedral, the roof of the Morning Chapel.

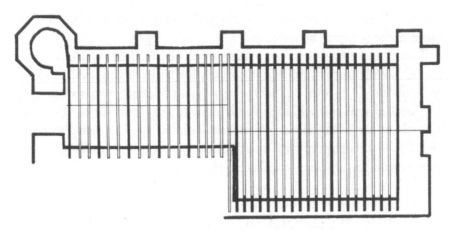

13. Lincoln Cathedral, plan of the high-roof of the Morning Chapel.

Cressing (Hewett 1980, 105), and these Chichester specimens are therefore likely to date from the later 13th century.

Lincoln Cathedral, the Morning Chapel

'Situated at the north-west of the cathedral, it is generally held to date
from about 1220-30, along with the outer and upper (Gothic) parts of
the west front' (Dr. J. H. Harvey, personal communication, 1 February
1981). Like the Consistory Court and the Bellringers' Chapel one of
the Norman towers reduces its span, enforcing a roof with two ridge
heights and widths; this is illustrated in Figs. 12 and 13. Both parts of
this roof have been repaired upon several occasions, but enough survives unaltered to enable their original characteristics to be assessed. As
indicated by the plan the bay-length in the largest part of the roof was
of *three* common-couples between the tie-beams, of which there are six.
The original couples and tie-beams have a single and central wall-plate,
which is cross-halved and continuous, but later repairs have placed two
wall-plates, passed through full-size trenches, in the north-eastern and
smaller of the roofs; and saltire braces lying in the horizontal plane
have been fitted to the feet of 21 couples at the south-west. At another,
unknown, date collar-purlins of pine were fitted over both upper and

37

lower collars by means of cross-cogged joints, doubtless inspired by those in the roof of the nave.

The transverse design was that which seems basic to the Lincoln roofs, having upper and lower collars, both chase-tenoned, soulaces with notched lap joints of archaic entry, and raking-struts with similar notched laps to their tops and barefaced lap-dovetails to their feet. The oldest surviving ashlar-pieces have notched lap top-joints that may be taken to be original, whilst others dating from various repairs have either chase-tenons, or are spiked to the rafters.

Lincoln Cathedral, the Chapter House Vestibule

'Both the documentation and the architectural detail place the Chapter House and its timber roof not before 1220, nor after 1230 at the very latest' (Dr. J. H. Harvey, personal communication, 1 February 1981). The roofs to this vestibule are not necessarily one and the same design, but they do both employ soulaces with notched lap joints of archaic profile, themselves features having a curious distribution at Lincoln. The roof is illustrated in Fig. 14, where it can be seen that the design is for double collars with raking-struts between them, soulaces and ashlar-pieces, which last are variously tenoned or nailed, presumably at different building dates. The tie-beams used do not relate to wall-plates, so far as may be determined today. The shortest bays contain two couples, and the longest contain three common-couples – in the light of which some rebuilding is postulated. The apexes of the rafter-couples are halved together, and the raking-struts and collars are chase-tenoned, all of re-used timber.

Lincoln Cathedral, the Consistory Court and Bellringers' Chapel

'Near 1225-35' (Dr. J. H. Harvey, personal communication, 1 February 1981). The roof over the Consistory Court is pitched at the steep angle characterising Lincoln, and framed into seven cants, having notched lap joints to its soulaces and upper collars, clearly designed as a roof intended to contain its own tendencies to spread laterally. At the centre of the rectangular vault-cell it covers, two tie-beams were fitted, with two common-couples between them, as shown on the plan, Fig. 15.

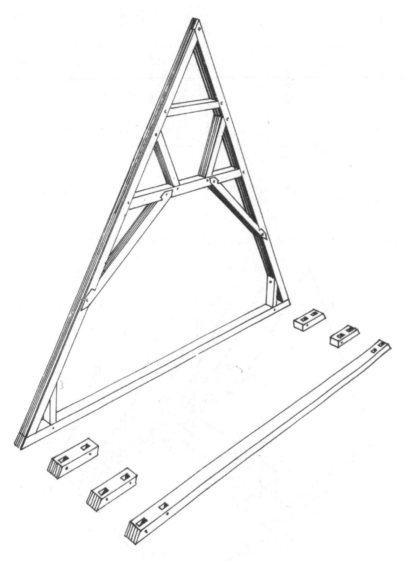

14. Lincoln Cathedral, the roof of the Chapter House vestibule.

They were stilted in the elaborate manner illustrated in Fig. 16, and in this form they clear the high crown of the domical Anjou-type vault, which they must have failed to do with their given eaves and parapet heights had they been straight and continuous beams. In addition to this degree of specialised design the southern pitch of this roof, which is

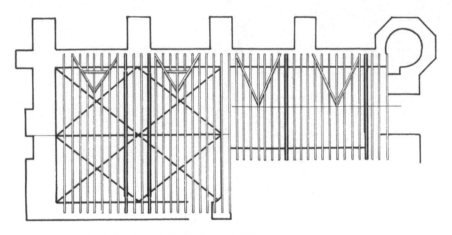

15. Lincoln Cathedral, plan of the high-roof of the Consistory Court.

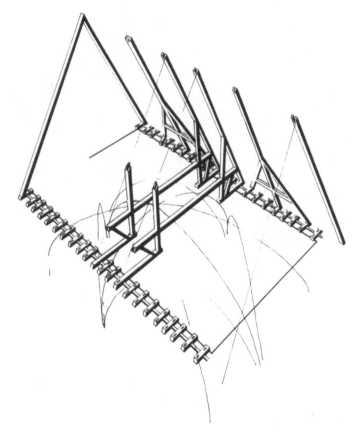

16. Lincoln Cathedral, the high-roof of the Consistory Court, showing the relationship with the vault beneath.

the only one externally exposed, is provided with the best anti-racking device known in the whole context of cathedral roofing – bifurcated couples which are, in effect, sole-braced in pitch on both sides of their bases. In these cases the principal-rafters are in one piece as low as to the short side-purlins, connecting the two inclined legs; these have countersunk forelockbolts at their junctions. The details of design and construction are illustrated in Fig. 17, showing the use on the same

17. Lincoln Cathedral, the high-roof of the Consistory Court, showing the higher rafter-couples.

components of both raking-struts and raking-ties outside them, of notched lap joints with refined entries, and barefaced lap-dovetails – the last twice pegged. It must be emphasized that all notched lap joints in this roof have refined entry angles. The wall-plating largely survives, and is of the type used for the late 11th-century roof at Chipping Ongar, Essex, in which a slender square-sectioned fillet served to align the sole-plates, fitted by means of cross-halvings (Hewett 1982, 3-4). Fig. 16 shows the relationship of the roof to the vault beneath it, and Fig. 18 shows a longitudinal section through both the chapels in which the reduction of ridge-heights is evident. This was necessitated by the retention of the western towers of the cathedral built by Remigius in 1073. By this a constant pitch was maintained, and off-set to the south, where a diagonally halved gable was built of studwork effecting the transition. The specialised nature of the lower rafter-couples is shown in Fig. 19, which underlines the uncommon degree to which this novel structural concept was studied. The collars either side of the rack-proof couples, virtually tripods, were raised; and for that reason fitted with extra long soulaces in the northern pitches.

This roofing system is probably the only one surviving in an English cathedral that can be proved to relate to the stone vault beneath it. As such it should be the most closely-dated example we have. 'This method

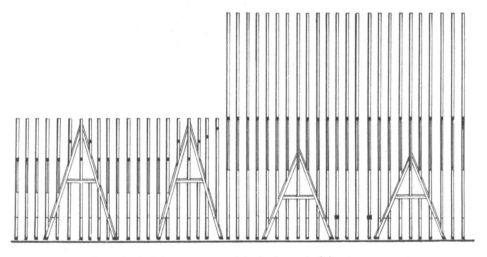

18. Lincoln Cathedral, long section of the high-roof of the Consistory Court.

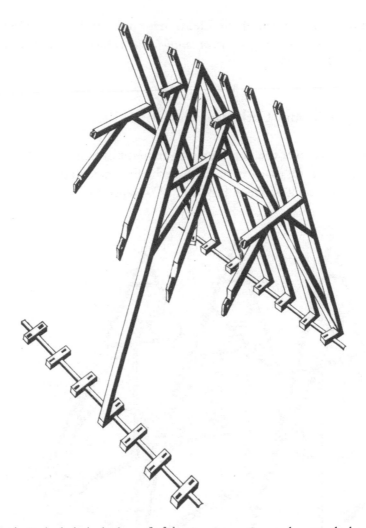

19. Lincoln Cathedral, the high-roof of the Consistory Court, showing the lower rafter couples.

of vaulting, characterised by a close network of ribs and by small penetrations, originated in Anjou in the early 13th century, but followed a complex pattern of evolution in which England played an important part. Three main types have to be distinguished: (a) Domical vaults on square plan, with short orthogonal penetrations. These are common in Anjou from *c.*1210: Saint-Florent, near Saumur, La Toussaint and Saint-Serge at Angers, etc. Similar vaults, but with a flatter curve, are found over the lantern of the Liebfrauenkirche at Trier by 1253 and *a little before in the Consistory Court or south-western chapel at Lincoln Cathedral*' (Focillon 1963, 151).

Salisbury Cathedral, the North-East Transept

The high-roof over the north-eastern transept of Salisbury Cathedral is the original one for that arm of the church, illustrated in Fig. 20. The eastern arms were built between 1225 and 1237, when Master Nicholas of Ely

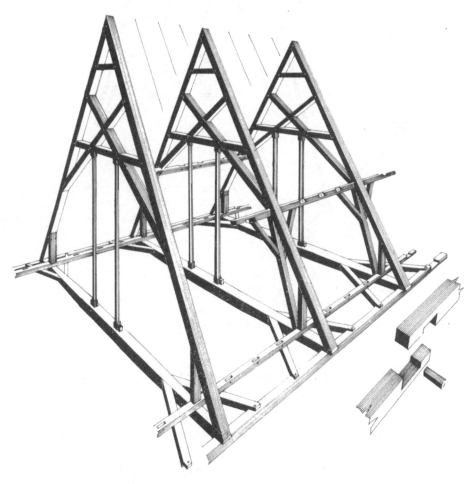

20. Salisbury Cathedral, the high-roof of the north-east transept.

was the architect, and the name of Master Elias of Dereham is somewhat intangibly associated with the fabric (Harvey 1974, 239). The elegance and highly-advanced structural design of this work place it, immediately, among the very few survivals left to us from the 'great age' (Rigold 1975, 431). Among the most striking developments is the use of carefully cross-sectioned tie-beams which have their volumes increased towards their centres – the point of maximum extension stress – and branching ends that extend their action to every third of each bay's length. Racking was rendered impossible, under all normal contingencies, by side-purlins that were arristrenched to locate the rafters, supported by in-canted prick posts, the latter arcaded to ensure their vertical stability. Scissor-braces were also used, in principle, but were here built from five components to resist mainly compressive stress. Tall, elegantly-chamfered queen-posts were, despite their slenderness, designed to support the collars – again in compression – and secret notched lap joints appear to have secured the tie-beams' branches to the heavy wall-plates. The scarf joint used for the latter is very rare and is only known elsewhere at Winchester Cathedral. There are grounds for assuming it to have Middle Eastern origins (C. L. Striker, personal communication, 1975).

Boxgrove Priory, Sussex

The great church of Boxgrove Priory survives as the parish church in a very complete state. It was founded as a cell of the Abbey of Lessay in the Cotentin peninsula, by Robert de la Haye, Lord of Halnacre in c.1115. There were three main periods of building activity, from c.1120, c.1170 and c.1220 – the last being, of course, in the Early English style and relating to the eastern bays of the choir (Ratcliff 1972, 7). The existing roofs are of great interest. The high-roof ranging eastward from the crossing tower is built in 11 bays of varying lengths and 'builds'. It appears that the two eastern bays were the final achievement, and these comprise five and six common-rafters each, framed in seven cants with double collars, chase-mortised together at all points. A collar-purlin was mounted on three side-braced crown-posts of which the western and eastern ones are terminal to the build, and have only one purlin-brace. The eastern bay is illustrated in Fig. 21. The bays westward from

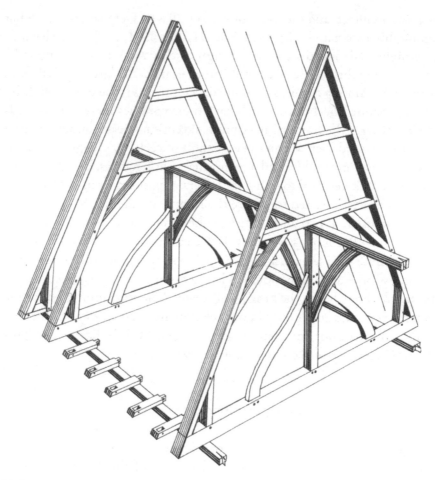

21. Boxgrove Priory, Sussex, the eastern bay of the high-roof.

these were reinforced by intruded crown-posts and collar-purlins at some date after their use in the final bays. The early date of the eight bays west from the tower is confirmed by their being but three commonrafters in length, and originally framed as of seven cants.

Romsey Abbey, Hampshire (*Fig. 22*)

Some distance south-west from the abbey church, in what is now a private house, survives the timber roof of what is thought to have been the

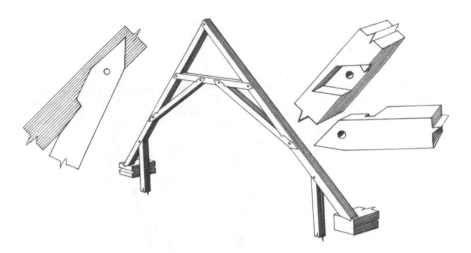

22. Romsey Abbey, Hampshire. One rafter-couple of the roof the refectory.

refectory. This is a design of seven cants, probably dating from *c.*1230, and assembled by notched lap joints. The soulaces have *open* notches at both ends, but the collars have *secret* notches. This is remarkable in that the strongest, but most expensive, form was used judiciously and only at the point judged to be most severely stressed (see Plate 4).

Lincoln Cathedral, the Nave (*Fig. 23*)

The eastern part of the present Nave was certainly damaged seriously at the fall of the Central Tower in 1237; the date 1238-55 includes not only the present lower stage of the Central Tower but also a good deal of reconstruction of the east end of the Nave, and must include substantial work on the high roof. The dates of *erection* of the Nave roof (the high roofs) must be *c.*1225-35 and (nearer the Tower) *c.*1240-55. The large bequest of timber of 1233 must surely have supplied an actual need of structural woodwork for the completion of the cathedral *as foreseen* in the time of Hugh de Wells (died 7 Feb 1235 — he had made his will two years before his death).

This is similar to those already described above the south-east transept and St Hugh's Choir. The span at this point exceeds 40 ft., and there is a total of 81 medieval rafter-couples, extending from the western face

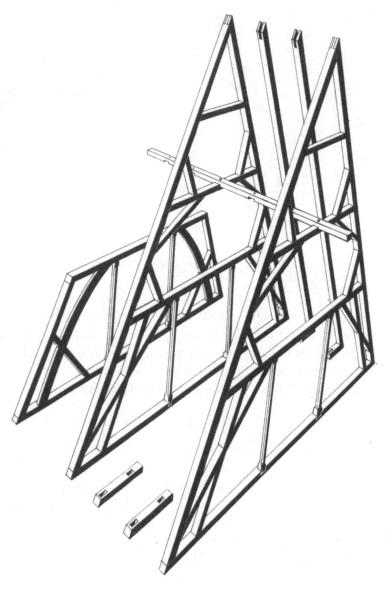

23. Lincoln Cathedral, the high-roof of the nave.

of the crossing tower to the eastern faces of the Norman towers that were retained from the foundation date. This roof is considered to date from between *c.*1225 and 1253, during the term of Master Alexander; yet it is alleged that 'the masters of the Lincoln workshop invented in the nave, not later than the 1230s, the rich star-shaped pattern of the tierceron vault' (Bony 1979, 44). The frames here have three collars, and both the lower ones were fitted with soulaces having notched lap joints of refined entry-angle. Truncated under-rafters were fitted also, and raking-struts above the first and second collars – whilst a collar-purlin was cross-cogged over the tops of the second collars, and carefully pegged through them. This component can be examined close to the western tower; if it is not strictly contemporary, it is at least as old as the occasion upon which the nave roof had been raised – from western tower to eastern gable. This last event does not appear to have been recorded in printed sources, but the complete flashing exists inside the western gable, where it only just misses the lancets – and appears to contain fragments of lead. The truncated rafter-couples that recur throughout the nave also occur in St Hugh's Choir. Insofar as it has been possible to examine the rafters of this roof they have been found to have been scarfed with the splayed joints noted in the roof of the Angel Choir.

Wherever checked (a huge task) the entry-angles of the soulace notched laps are refined. This places them in the later 12th century, like the majority of those at Wells. They may also date until the Interdict of 1209-13, as Professor Bony's observation indicates, and re-use of existing cathedral timbers is essential to account for this. A further timber profile for a rafter-couple exists against the western face of the crossing tower. It may relate to the rebuilding of the nave-roof which was undertaken at some unknown date before the completion of the existing west gable, which still has an internal flashing, apparently with lead *in situ*, and substantially lower than the present roof-line. The jointing of this couple is scarfed and forelock-bolted, in the manner to be found in both nave and Angel Choir rafters, of the splayed and sallied 13th-century type; the pegging is effected in both edge and face-planes. It is considered that earlier rafters were re-assembled by means of the two 'jigs' found to east and west of the central tower, their upper limits being extended by scarfing. It must be stated that the scarfing of rafters is rare; extremely long rafters were grown for Westminster Abbey. If the scarfs of the Lincoln rafters were effected merely from necessity, we should not find them consistently placed at a height that is precisely uniform, as is there the case.

Beverley Minster, Yorkshire (*Fig. 24*)

'On the 20th of September 1188, as more than one medieval chroni-
cler tersely notices, the town of Beverley was burnt "together with
the noble church of Blessed John the Archbishop". Between 1220
and *c.*1260 the present choir, the preter and lesser transepts and the
now vanished chapter house north of the choir, were completed,
the whole being a superlative essay in structural elegance and design

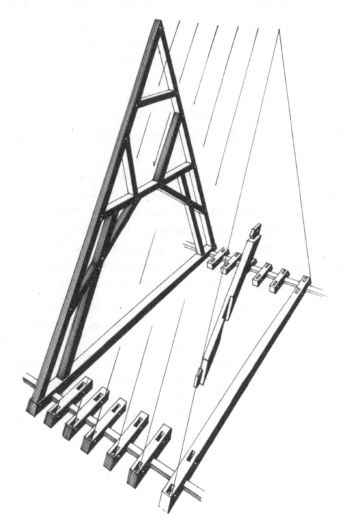

24. Beverley Minster, the high-roof of the nave.

... The building programme must have been well advanced when, in 1252, Geoffrey de Langley, Justiciar of Forests, was ordered to supply forty oaks from Sherwood Forest as a royal gift to the works. The name of the master mason who conceived this great design (which determined the size and character of the new church) is not known. But any surmise as to identification of those in charge of the work by, say, 1240, could not fail to include the name of Robert de Beverley who became the King's Master Mason in 1260' (Macmahon 1970, 8).

Greyfriars, Lincoln

About 1237 the Grey Friars began to build their church at Lincoln (Martin 1935, 44), and the first completed phase of their building was roofed with 'compass' timber as illustrated in Fig. 25. This oldest part of the roof was completed, it is believed, by *c.*1260. With two collars, considered to act in extension since notch lap-jointed, it was framed into a semi-circular archivolt by four curved timbers – soulaces and ashlar-pieces. The jointing of these last items tends to evade definition, but the laps constitute dovetails, and an exploded inset illustrates their diminution, which although unproved is essential to preserve their strength. The face-pegging of all the lapped components is to prevent their sidewise dislocation.

Westminster Abbey, the North Transept

Documentary evidence relating to the completion date of the three eastern arms of the Abbey has been published, and 1259 is most probable; the master carpenter to Henry III, Master Alexander, directed the work (Colvin 1963, I, 133, 144). The roof comprised 36 frames, each with two collars and scissor braces, the feet of the latter being fitted into secret notched lap joints. The tie-beams were placed at intervals of eight couples inclusive, and were of the distinctive type with branching ends. The ashlar pieces were in-canted. Perhaps the most interesting feature of the roof was the fitting of diagonal braces into trenches on the outer

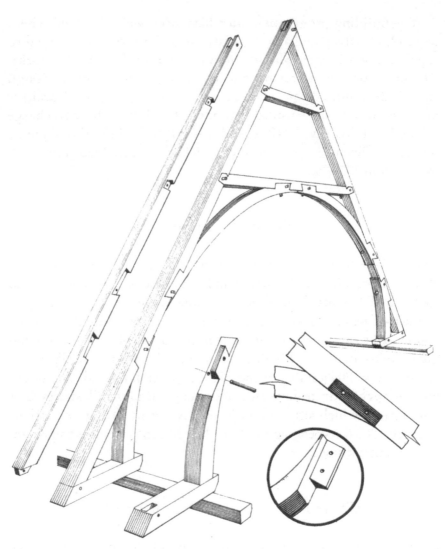

25. Greyfriars, Lincoln the eastern part of the nave roof.

faces of the rafters, at the ends of its range; these were to inhibit the
racking of the couples. A short section of this is illustrated in Fig. 26.
The crown-post standing centrally on the tie-beam rises to the height
of the crossing of the scissor-braces. Nothing remains today, because all
but the eastern apse was replaced shortly after 1964.

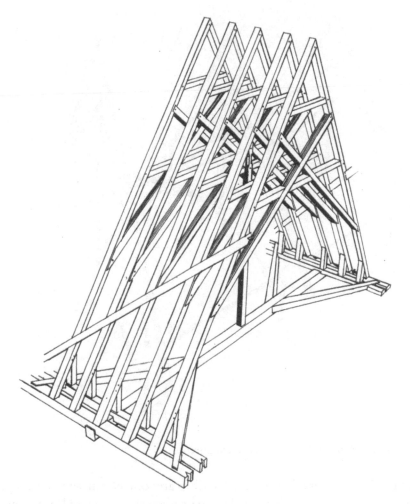

26. Westminster Abbey, the high-roof of the north transept.

Salisbury Cathedral, the North Porch (*Fig. 27*)

'All the roofs at Salisbury were probably completed before the west front, i.e. before 1266, and the building of the nave westward from the crossing occupied the years between 1237 and 1258; the porch, there-fore, may have been roofed c.1250' (Harvey 1972, 157). However, a remarkable introduction was made in the form of crown-posts, with-out collar-purlins, which were saltire-braced – a device designed to

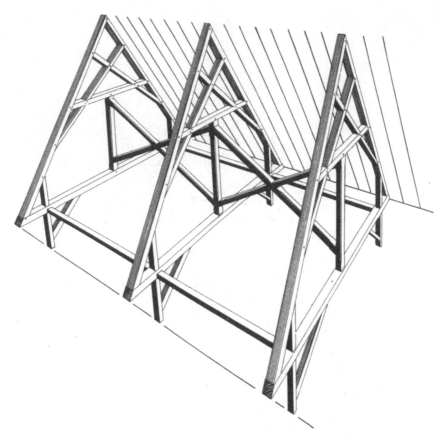

27. Salisbury Cathedral, roof of the north porch.

contain the racking of the couples within any one bay in which it might commence. Strangely, this development seems to have had little effect, and no comparable or derived crown-post roofs are known at the time of writing. Additionally, this roof had compressive scissor braces on every couple, the type apparently incepted when the eastern arms of Salisbury were roofed.

Lincoln Cathedral, the Angel Choir

'Angel Choir – these roofs cannot have been *designed* before 1256, as the Angel Choir is a wide span where there was no previous wide building to get roofs from (nor, indeed, any narrower building in this area)' (Dr. J. H. Harvey, personal communication, 1 February 1981).

The architect was Master Simon of Thirsk (Harvey 1974, 231). The roof must have been completed many years before 1280 because, as was necessary and invariable in such cases, the vaults were built inside the finished roof-space. One bay of this is illustrated in Fig. 28. The design relates closely to that of the nave roof, and has slender

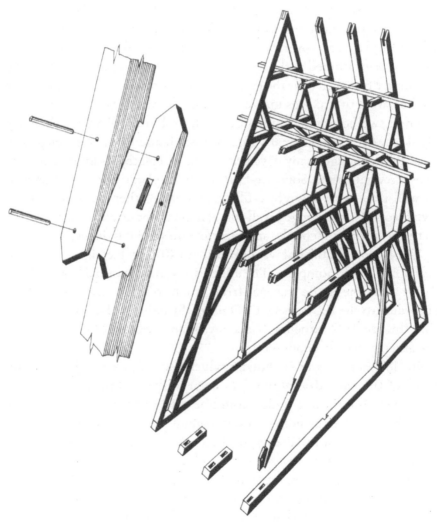

28. Lincoln Cathedral, the high-roof of the Angel Choir.

octagonally-sectioned posts acting as supports to the lowest of three collars in each frame. Open notched laps were used for the collar-tying soulaces, and chase-tenons for the struts above the collars. The rafters are over 50 ft. long, and all were scarfed together with the splayed and sallied joint shown in the drawing, the splay diminishing down the outer face. Racking was inhibited by the provision of *three* collar-purlins, each one trenched across its underside to house the successive collars.

This roof is not only a design derived from those of St Hugh's Choir and the nave; it develops both, by the addition of two collar-purlins which are placed one over two, above the second and third collars. These can best be examined at the eastern gable, where it can be seen that the purlins are trenched over the collars; the pegging alone differs from that of the nave, but the pegs may occupy blind holes. The bay-lengths are again of two common-couples between tie-beams, of which there are eighteen. All were originally queen-posted, and those posts elegantly chamfered; though many have been crudely replaced, original ones survive at several points. Some tie-beams remain, placed upon short lengths of wall-plate with hewn face-fillets, and some others are mounted on two pieces of cross-cogged wall-plate; none of these items reach as far as to the next adjacent couple, and therefore no system is deducible.

The Norman cathedral was begun about 1072 by Remigius, and took about 20 years to complete. The dimensions were; internal length about 310 ft., and the single transept from north to south just over 122 ft., according to the Rt. Rev. D. C. Dunlop. In addition, Dr. Harvey states that this pre-1192 cathedral had a nave span of about 28 ft., and across the aisles internally about 66 ft.

In 1185 there was an earthquake which was severe enough to destroy most of the cathedral which had been built by Remigius. All that remained of any value which could be used in conjunction with the new church were the north-west front and the west towers. This new church was begun in 1192 by St Hugh of Lincoln. The original timbers of the roof were re-used for the church. For example, the Angel Choir and the nave were much higher than the original. What appears to be Norman walling and arcading exist in the north triforium of this choir, suggesting that some part of the Norman wall was there retained and refaced on the south; the width of the cathedral being

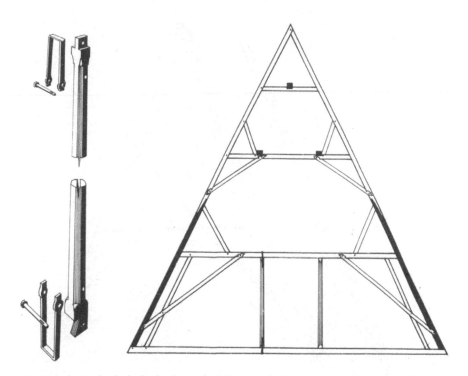

29. Lincoln Cathedral, the high-roof of the Angel Choir showing the re-used rafters in black (not to scale).

increased further to the south, and a new southern wall built *c.*1192. Fig. 29 (not to scale) illustrates the cross section of the new roof built for the Angel Choir. Since the roof timbers available to him were less than 40 years old at that time they were re-used, together with their earliest joints – such timber was certainly too good and expensive to waste. Fig. 29 also illustrates precisely why the scarf joints must occur where they do in fact occur. On both sides of the drawing the sections that were retained are shown in black. This allows for a relatively rapid rate of rebuilding at this time.

Greyfriars, Lincoln (*Fig. 30*)

This roof is a continuation of the one mentioned previously at Greyfriars. Its progress was interrupted but in this instance the date of

30. Greyfriars, Lincoln, the western part of the roof.

the interruption is less certain; it is thought to have been *c.*1260. This break is of much interest, for the same reason as the one at Wells, that the type of construction changed during the interval. The roof-couples belonging to the second phase are scissor-braced and straight, but still jointed with notched laps. There are 28 feet of this second roof comprising 17 couples, which can be ascribed to the third quarter of the 13th century (Martin 1935, 44).

Wells Cathedral, the Choir

The choir of Wells Cathedral was extended eastward at some date prior to the conversion from eaves to parapets, which work is believed to have been completed by 1310 (L. S. Colchester, personal communication, 1972). One principal-rafter couple from each of the two 'builds' is shown in Fig. 31, in which the eaves triangles can be seen to have been depleted, and tie-beams bolted across to stabilise the resultant forked rafter's feet. These tie-beams are rare in that they are scarfed, with the splayed and tabled joint shown enlarged at right – a joint designed specifically to resist extension stresses. Despite the very complete documentation of Wells the date of the choir roof cannot be established, only the fact that it must precede the conversion to parapets which caused its basal modifications. The evidence of typological comparisons, however, makes it impossible to assume that the design relates to the original choir of *c.*1175 (Harvey 1972, 163).

The Old Deanery, Salisbury (*Fig. 32*)

The top-plates of the building incorporate what may be the earliest example of splayed scarfs with tonguing, and the work is datable by a deed of gift issued by Robert de Wykehampton, who became Bishop of Sarum in 1253. This document states: 'we assign, grant and give forever those our houses in the Close of Sarum, which we were accustomed to inhabit when Dean'. The evidence suggests a date of construction

31. Wells Cathedral, the high-roof of the choir.

between 1258 and 1274 (Drinkwater 1964, 44). This scarf unites the lengths of top-plate and may be defined as stop-splayed with tabling and square under-squinted abutments, tongued and grooved part-length, laterally keyed and with one edge-peg.

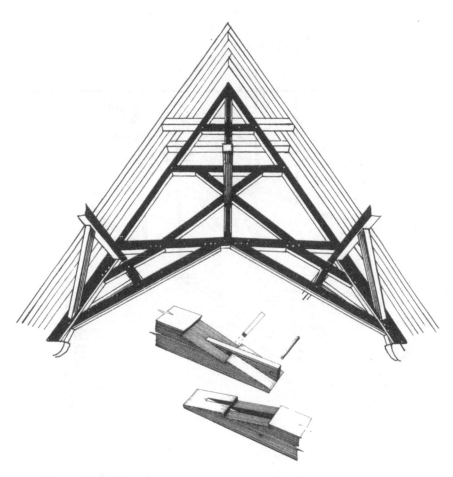

32. Salisbury, the Old Deanery.

Tewkesbury Abbey, the North Transept (*Fig. 33*)

A remarkably elaborate and eclectic roof system survives here, and its survival testifies to its efficiency; the dating of the sub-structure does not assist with that of the carpentry but technological details are comparable with the roof of the Wells retro-choir of *c*.1329-1345. Three bays exist, each having seven couples within a distance of 13 feet. The tie-beams are straight, and suspended at their centres by queen posts which have lap-dovetailed ends at top and bottom. These queen posts

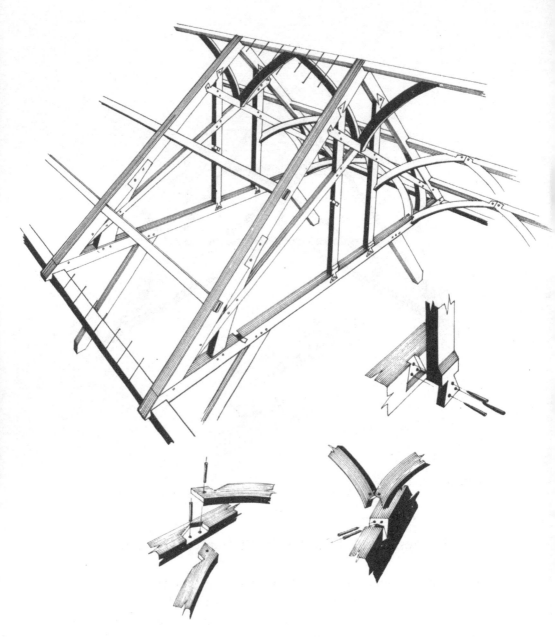

33. Tewkesbury Abbey, the high-roof of the north transept.

are scissored, as shown in the drawing, and compassed struts resembling ashlar pieces are jointed into the assembly. A ridge-piece is socketed into the apexes of the principal rafters, and likewise two side purlins into each slope. The ridge and purlins are elaborately braced with curved timbers that spring from appropriate, but most unusual, points in the assembly – those to the ridge, for example, being stepped in bird-mouth joints worked at the crossings of the scissor braces. Others, as the drawing shows, are stepped in similar sockets cut into either the queen posts or the tie-beams.

Gloucester, Blackfriars, the Nave

Blackfriars at Gloucester was built as a church, having become estab-lished there in 1239. Royal grants cover the years 1241 until *c*.1267, by which date it was probably completed (Rackham, Blair and Munby 1978, 105 and 120-1). The roof shown in Fig. 34 is that over the nave; it is scissor-braced with open notched laps at the feet of the scissors, and it has an elaborate system of wall-plates, three on each wall. Perhaps the most important feature of the roof is the wind-bracing, which is spiked to the soffits of the rafters; the spike-heads are driven into shal-low square counter-sinkings, which technique relates them to the spikes of the west doors of Salisbury Cathedral and Westminster Abbey. They are clear evidence that carpenters were showing some concern over the

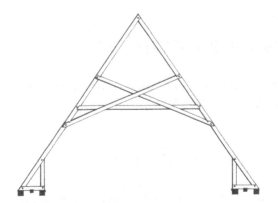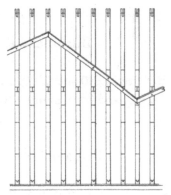

34. Gloucester, Blackfriars, the roof of the nave.

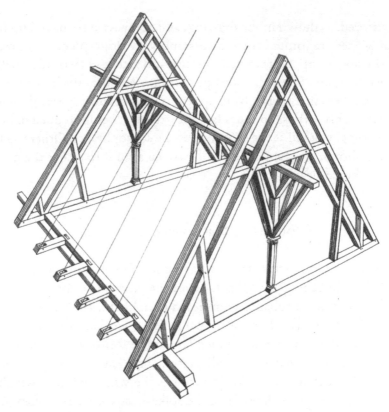

35. Bushmead Priory, Bedfordshire.

problems of roofs 'racking' along the lines of their lengths during the 13th century.

Bushmead Priory, Bedfordshire (*Fig. 35*)

'Between 1280 and 1300. Large arched braces became common in western England from about 1300' (Bony 1979, 83).

Bisham Abbey, Berkshire, the Hall (*Fig. 36*)

The building known as Bisham Abbey, on the Berkshire side of the Thames near Marlow, is in fact a 13th-century manor house built

36. Bisham Abbey, Berkshire, the roof of the hall.

by the Knights Templar (V.C.H. III, 139), ascribed to *c*.1280. This conforms to the H-plan and has a ground-floor hall of remarkable size – measuring 52 ft. by 33 ft. 6 ins – roofed with 29 scissor-braced rafter-couples. At the western end a tie-beam passes under the double wall-plates and mounts a tall crown-post with scroll-moulded capital, with straight braces at each side to the collar. Two empty mortises indicate the previous existence of a butted collar-purlin and its brace, suggesting the former existence of several such crown-posts. The internal wall-plate is also moulded, as illustrated (Fletcher and Hewett 1969, 220).

Exeter Cathedral, the Presbytery

The high-roof above the presbytery of Exeter Cathedral, of which one rafter-couple is illustrated in Fig. 37, is the oldest portion of timber roofing surviving in the cathedral. The building of the presbytery was begun in 1288 from its eastern end and moved westward, and was one

37. Exeter Cathedral, the high-roof of the presbytery.

bay into the nave by 1317 (Harvey 1972); the eastern part of the present church must date from *c*.1290. These couples have scissor-braces with open and archaic notched laps at their feet, and small raking-struts from scissor to rafter for the support of side-purlins. The wall-plates were inserted as distance-pieces between the sole-plates as the work progressed. It cannot be demonstrated from the jointing that the purlins were intended to inhibit racking.

Winchester Cathedral, the Nave

The high-roof over the eastern end of the nave of Winchester Cathedral cannot be clearly associated with any of the documented building dates yet known; it is illustrated in its present and altered condition in Fig. 38.

38. Winchester Cathedral, the high-roof of the nave.

The one detail that assists with a date ascription is the scarf used for the inner of the two wall-plates, which is shown at right. This is the only example known of the kind besides those at Salisbury (Fig. 102) and a relatively close date is probable.

Winchester Cathedral, the South Transept

The south transept of Winchester has an early roof that has been little altered, which is shown in Fig. 39. It is framed of massive oak and chase-tenoned. It cannot be dated by written records and does not pertain

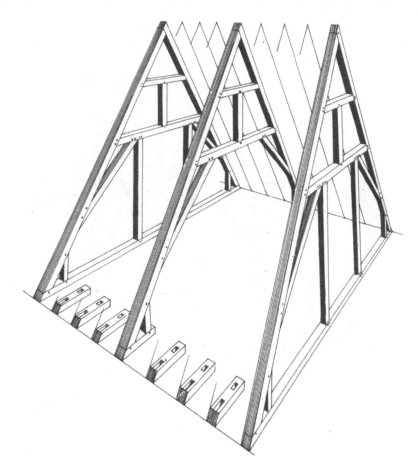

39. Winchester Cathedral, the high-roof of the south transept.

to the building of the transept. The bays are short, comprising only three common-rafters each, and this fact, together with the design of a mountant between two collars above a seven-canted void, all chase-tenoned, allows of a date close to *c*.1300.

Wells Cathedral, the South Choir Transept

The south choir transept at Wells is roofed with the relatively small and simple structure shown in Fig. 40, framed sometime before 1330, when the transept was dedicated. It comprises 10 couples of collared rafters forming a single bay with a tie-beam at each end; curved angle-braces were fitted at each corner, secured with notched lap joints as illustrated. The acute angles between these subtly-curved braces and the tie-beams are considered to represent the final development of the branching tie-beams described earlier in 13th-century contexts.

The Chequer, Vicars' Close, Wells

The dates of the various buildings of Vicars' Close are problematical, although three documents exist which suggest that a fair number were finished and in use by 1348. These documents are the will of Alice Swansee, dated 7 November 1348, a licence in mortmain dated 3 December 1348, and a deed confirmed by the Dean and Chapter of Wells on 3 January 1348/9. This last refers specifically to 'the Vicars living in the new building erected by him and eating together in the common hall of the said building', which leaves the dates of the other buildings and of the vicars' houses uncertain. The Chequer leads north from the hall, at the same level, connecting it to the tower standing at the northern end which embodies parts (buttresses and arches at ground level) of an earlier tower of the Decorated period. The roof is in five bays with braced collars and side purlins that have bracing beneath them and inverted bracing above them, as shown in Fig. 41.

According to Mr. L. S. Colchester, formerly Cathedral Secretary of Wells, 'Favoured by a Wells carpenter between 1420-50, and almost

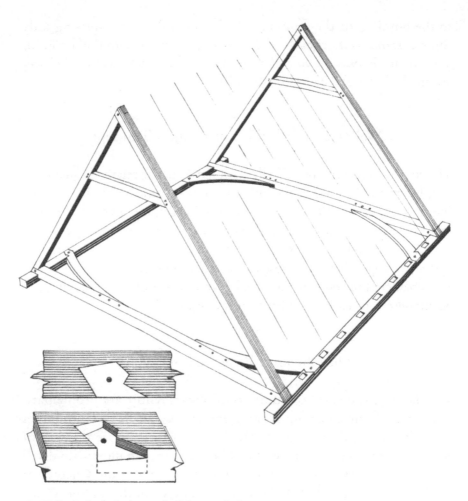

40. Wells Cathedral, the roof of the south choir transept.

certainly derived from the "inverted arches" under the Cathedral tower'
(personal communication).

Exeter Cathedral, the Nave

The high-roof of the nave is dated between *c.*1325 and 1342, when
canvas was purchased to block the great west window, an unnecessary act had no roof been in place. The architect throughout the

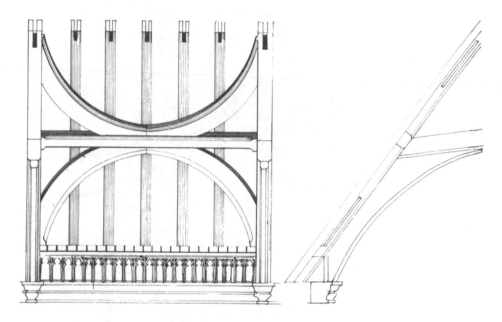

41. Roof of The Chequer, Vicars' Close, Wells.

period was Master Thomas Witney, 'one of the few proven examples of an architect who was both mason and carpenter, and an outstandingly advanced designer' (Harvey 1972). Witney's remarkable roof commenced at the second bay west of the crossing, where it joined on to the earlier one, but is divided into short bays of only three common-couples, and stabilised against racking by crown-posts with butted collar-purlins. During this operation 48 'great oaks' were purchased, among other things, confirming the visible implication of the design – that its frames were wrought in advance of the building's progress, and reared whenever a bay's length was completed. One such bay is illustrated in Fig. 42, which shows that the side-purlins of the earlier roof were continued, but now trapped between the collars, rafters and vertical struts, at the tops of which archaic open notched laps persisted in use. The scissor-braces were now exalted to the upper part of the rafter couple, and chase-tenoned soulaces were fitted beneath the lower collars. The earlier distance-pieceing of the sole-plates was continued.

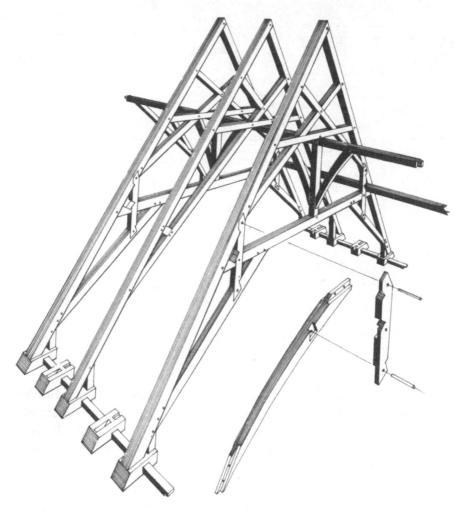

42. Exeter Cathedral, the high-roof of the nave.

Exeter Cathedral, the Transepts

The two transeptal roofs, connecting the Norman towers, were framed by Thomas of Witney in almost the same way as that of the nave, but with one innovation in his lap-jointing – a counter-sallied cross-halving, which is encircled on the drawing, Fig. 43. These two were valleyed onto the nave roof by valley-boards, as illustrated. It is possible

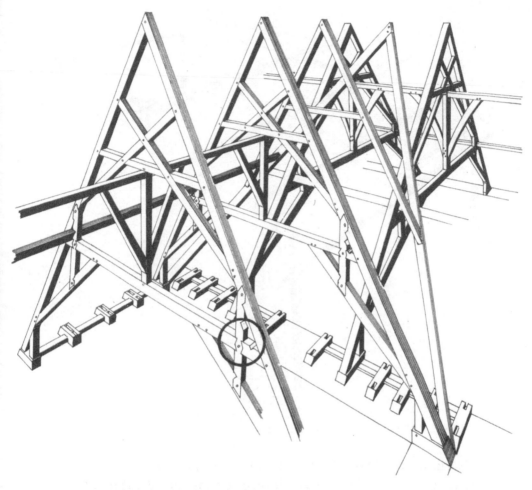

43. Exeter Cathedral, the high-roofs of the transepts.

that Witney's change of both the situation and the function of scissor-braces preceded the development described in respect of Salisbury's transept, in view of the relationship of their known dates.

Bristol Cathedral, the Choir

The high-roof of the choir of Bristol Cathedral survives in sufficient detail to produce the drawing, Fig. 44. This part of the church

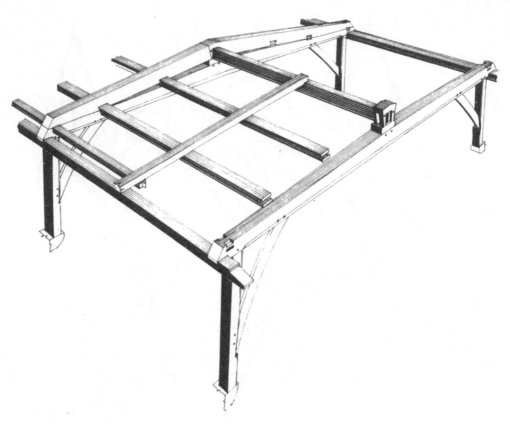

44. Bristol Cathedral, high-roof of the choir.

is ascribed to *c*.1311-1340 (Harvey 1974, 212), and was the work of a master of unparalleled genius whose name has eluded all records. No joints were used in its framing that preclude the possibility of the earlier date ascription. The design is of great importance, being the precursor of the camber-beam roofs that were expounded during the Perpendicular period; and the main frames possessed many of the assets of the 'built' beams that were advocated after that period. The low pitch is also important and is in clear contrast to the majority of early medieval roofs. As illustrated, the lower of the two butted side-purlins is fitted by pairs of single-tenons, and each of them penetrates a different timber; the upper one enters the principal-rafter and the lower enters the tie-beam, locking them together to inhibit flexibility! This same effect of rigidity was sought, centuries later, by fitting blocks into trenches cut across the adjacent faces of two bridging-joists which were subsequently bolted together. The dwarf 'king-posts' have sunken faces to accommodate the spurred shoulders of the massive ridge timbers, and their edges provide bearings for the principal-rafters. Five common-rafters were fitted in each bay; it has not proved possible to determine the jointing at their apexes.

Bisham Abbey, the East Range (Montagu's Great Chamber)

The building of this range, which incorporates a cloister and is stone-built, is believed to have taken place after 1336, when William, Lord

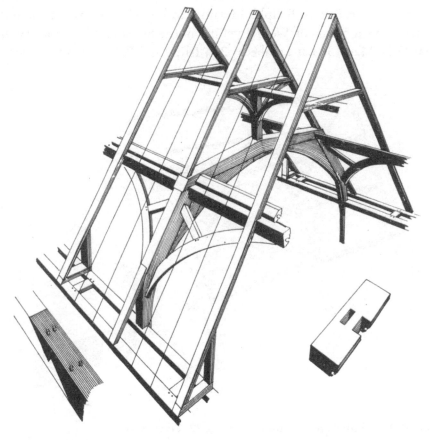

45. Bisham Abbey, Berkshire, the roof of the east range (Montagu's Great Chamber).

Montagu (1301-44) obtained possession of the manor (Fletcher and Hewett 1969, 222). The mouldings worked on the side purlins confirm this dating of the work and are shown in Fig. 45, which also includes the scarf joint used. The roof consists of five bays and is without base ties, having elbowed eaves blades secured with knees to the half-height tie-beams, on which a crown post and collar-purlin assembly is mounted. The internal span is 23 ft.

Wells Cathedral Precinct, no. 22 Vicars' Close (*Fig. 46*)

Although the dates of the vicars' houses are uncertain, as Mr. L. S. Colchester states: 'There is no reason to suppose that they were not completed in Bishop Ralph's lifetime. In his will he leaves corn and cattle "to the vicars of Wells dwelling in the houses built by me" (will dated 12 May 1363 at Lambeth, 244 Islip, in S.R.S. xix, 1903, 286)'.

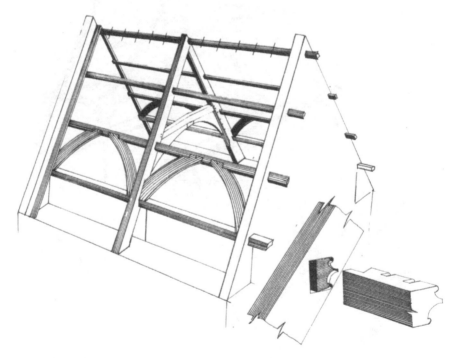

46. Wells Cathedral precinct, the roof of No. 22, Vicars' Close.

The design of the roof of No. 22 is common to all the houses, having principal rafter-couples with arch-tied collars, and three purlins in each pitch, with a ridge-piece. The middle purlins are wind-braced in-pitch. The lowest purlins are so set as to form internal eaves cornices. All the purlins are fitted into the principal rafters by means of tenons with soffit spurs, and in the case of the lowest ones this fitting was also scribed, as shown in the drawing. The arch-tying of these couples is also the earliest known example of that technique.

Winchester Cathedral, the Presbytery

A high-roof of great complexity surmounts the presbytery of Winchester Cathedral, illustrated by Fig. 47. This part of the building is dated to *c.*1315-60 (Harvey 1974, 243), when the architect responsible was Master Thomas of Witney, who died in 1356. There are no structural connections between the roof and the timber 'vaults' beneath it, but the roof was certainly designed to accommodate a vault since its frames have alternating low and high tie-beams – to stand in vault-pockets or to clear vault crowns. The wooden bosses of the existing vault are ascribed to the architect Thomas Berty (Harvey 1972, 165), and to the episcopacy of Bishop Fox (1501-28), being additions to an earlier vault.

Winchester Cathedral, the North Transept

The high-roof over the northern transept of Winchester Cathedral is illustrated in Fig. 48. This shows the forelock-bolted assembly of the paired queen-posts, which trap the scissored, or raking braces, which alternate in the succeeding frames. There is clear evidence for the design taking account of a vault beneath it, because some frames sit in the pockets, while the others clear the crowns. This roof relates to that over the presbytery (Fig. 47); a close dating of 1315-56 is proposed, and possibly the work of the same craftsman designer, Thomas of Witney.

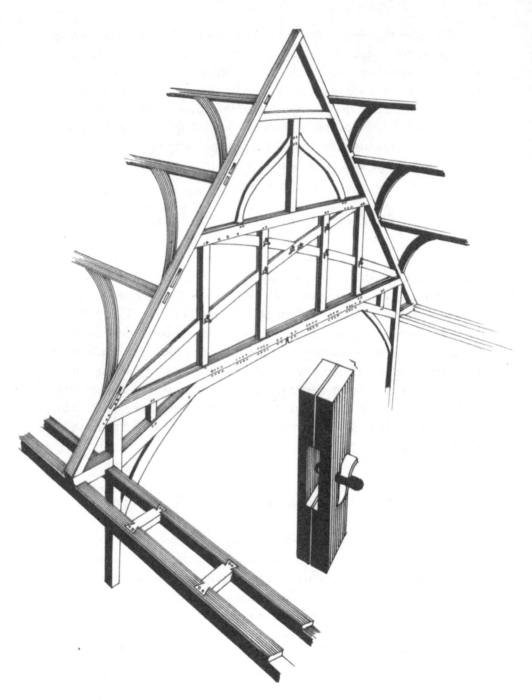

47. Winchester Cathedral, the high-roof of the presbytery.

48. Winchester Cathedral, the high-roof of the north transept.

Carlisle Cathedral, the Choir

The rafter-couple illustrated in Fig. 49 is one of the many identical frames comprising the high-roof above the choir of Carlisle Cathedral. This is dated with precision to the years 1363-95, and its architect was Master John Lewyn, during the episcopacy of Bishop Appleby (Harvey 1974, 221). A wooden ceiling is affixed to the archivolt of these frames, the bosses of which bear the armorial devices of the subscribers – which confirm the dating. Possibly the best of its type – the collar-arched category – it was prevented from any tendency to 'rack' by the integral nature of the attached and close-panelled ceiling. Only the apex

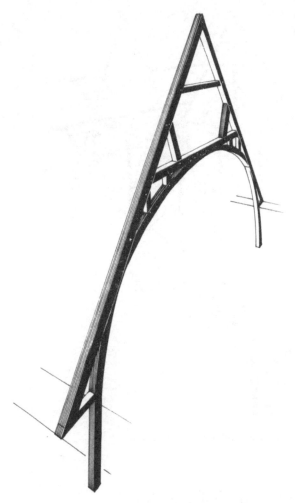

49. Carlisle Cathedral choir, one rafter couple.

region of this framing is examinable, and the eaves framing has yet to be established.

Worcester Cathedral, the Nave

The majority of the existing roof frames of Worcester's lengthy nave are as shown in Fig. 50. These were made of previously used oak-timber, probably from the original roofs of the late 12th and early 13th century. They are set stilted to clear the vault-crown, which does

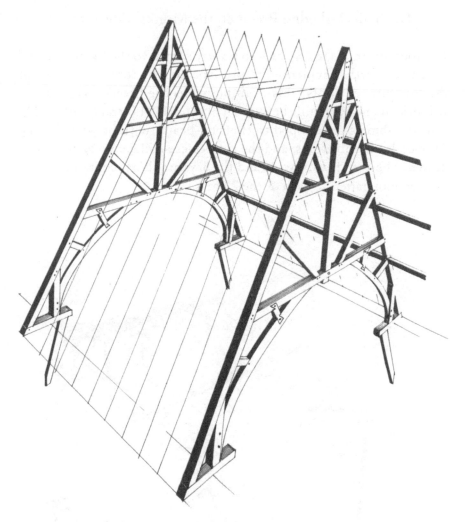

50. Worcester Cathedral, the high-roof of the nave.

not now rise above the eaves-level enough to justify this construction. None of the published information on the fabric assists with their dating, but numerous rafters are scarfed with the mid-13th-century joint, the splayed and tabled type with under-squinted butts, edge-keyed and face-pegged; whilst other re-used timbers in this and the eastern roof bear matrices illustrated in Fig. 213, indicating the type of previous Worcester roofs.

Durham Cathedral Precinct, the Monks' Dormitory

The roof illustrated in Fig. 51 is in no way similar to the last mentioned one, and is further evidence for an advocation by some carpenters of the low-pitched roofs; this example spans the Monks' Dormitory at Durham and was built between 1398 and 1404 (Stranks 1972, 20). The arch-braces are clasped to their tie-beams with box-ties having fleur-de-lys terminations, and the ridge-piece is elaborately cusped, as shown. The span is very wide and the bays are short, containing only five common rafters in each of the 23 bays.

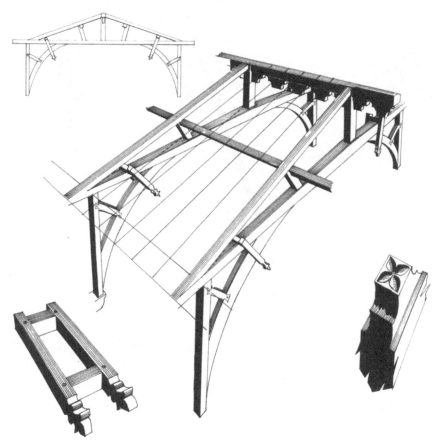

51. Durham Cathedral, the roof of the monks' dormitory.

Beeleigh Abbey, Essex

Beeleigh Abbey, dedicated to St Mary and St Nicholas, was re-founded near Maldon, in Essex, by Robert Mantell, *c.*1180, for Premonstratensian canons removed there from Great Parndon (R.C.H.M. II, 1921, 178-81). Of this, substantial parts survived the Dissolution and were converted into a residence, namely the Chapter House and the undercroft of the Dormitory. The Dormitory measures in plan 42 ft. by 21 ft., and its east wall is remodelled in 15th-century red brick (Pevsner 1965, 81), probably earlier than the residential conversion; and to this the impressive open timber roof must relate. Part of this is shown in Fig. 52. It was

52. Beeleigh Abbey, Essex, the roof of the dormitory.

framed in four bays, each comprising five common-rafters. Twin wall-plates were tied together at bay intervals where the principal frames were set, each having compassed soulaces and ashlar-pieces, and supporting short crown-posts and a collar-purlin.

Canterbury Cathedral, the North-West Transept

The north-western transept of Canterbury Cathedral has one of the few medieval roofs that survive there, illustrated in Fig. 53. The fabric

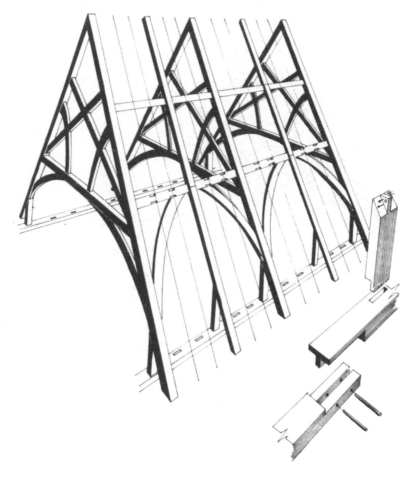

53. Canterbury Cathedral, the high-roof of the north-west transept.

of the transept was designed by Richard Beke between 1448 and 1455
(Harvey 1974, 21). This is a late development of the scissor-braced
category, in which the raking-struts actually scissor the scissors; all
the joints are compression-resistant, including those of the collar
arch-braces, which seems to indicate an error in design – unless those
chase-tenons have unseen modifications to resist extension. The scarf
used for the inner wall-plates is shown; this is of interest because it was

54. Peterborough Cathedral precinct, the roof of Table Hall.

used also for the monastic barn at Little Wymondley in Hertfordshire, which was carbon-dated by Professors Berger and Libby and gave one probability for 1475 (Berger and Libby 1967, 489). The same scarf was used in the great barn of Harmondsworth, Middlesex.

Peterborough Cathedral Precinct, Table Hall (*Fig. 54*)

This was the Hall in which the vicars choral ate together, as is recorded in the episcopal visitation of Bishop Scambler in 1567, when the canons were actually taking dinner in the Hall. This building has a ground storey of masonry, upon which is mounted a two-bay timber hall that is long-wall jettied. The principal couple has two collars, into the lower of which are tenoned two arched braces securing spur-ties – the *legender-Stühl*, or lying frame, of German carpentry. The first floor is framed with central tenons with housed soffit shoulders, and the bridging joist moulded with double ogees: the head of the ground-storey wall is cut to a great casement profile, with fleurons. These features together suggest an early to mid-15th century date.

Hereford Cathedral, the Vicars' Cloister

The cloisters of Hereford Cathedral were built in three stages; the roof illustrated in Fig. 55 covers the passage from the cathedral to the Vicars' College, dated to 1472 or a little later (Marshall 1951, 178). This is a plain king-post truss with remarkably deep-sectioned timbers carved in high relief; the wall-plate and side-purlins are moulded, as are the common-rafters, with profiles tending to confirm the date. The wall-plates are joined with the edge-halved and bridlebutted scarf illustrated.

Bath Abbey, the Choir

The choir of Bath Abbey has a high-roof constructed during the building period of the existing great church, between 1501 and 1539, of a complex design, as illustrated in Fig. 56 (Harvey 1974, 218). Originally

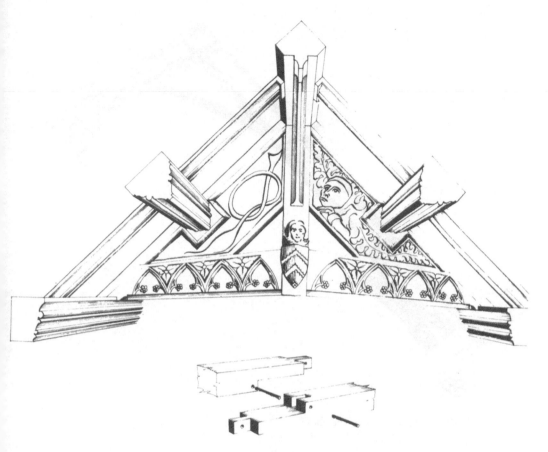

Fig. 55. Hereford Cathedral, the roof of the Vicars' Cloister.

curved and inverted wind-braces rose from the upper side-purlins to steady the principal frames, which were a curious hybrid formed with elements of earlier systems – scissor-braces, under-rafters and single hammer-beam. The frames are as illustrated with their lower side-purlins laid beside hammer-posts from which scissor-braces extend to the opposite principal-rafters. Humped collars complete the assembly, their end-joints designed to resist compression imparted by the weight concentrated on the raking-struts above them, and supporting the upper purlins. This is believed to have been an open roof, which now clears the existing vaults by less than head-room.

56. Bath Abbey, the roof of the choir.

King's College Chapel, Cambridge

King's College Chapel, Cambridge, was roofed as it was built, in two stages with a considerable time-lapse intervening. Fig. 57 could equally well represent a bay from either building operation; because both are visually identical and their joints have to be probed to establish their technological dissimilarity. It is common to every known break in the continuity of building operations, such as at the Wells nave, the Wells choir and the Lincoln Greyfriars' church, that changes in structural design or developments in jointing occurred during the interval when building was suspended; and at King's Master Richard Russell introduced the tenon with diminished haunch in the second 'build'. The roof itself was of a relatively poor design and arched to its collars, with

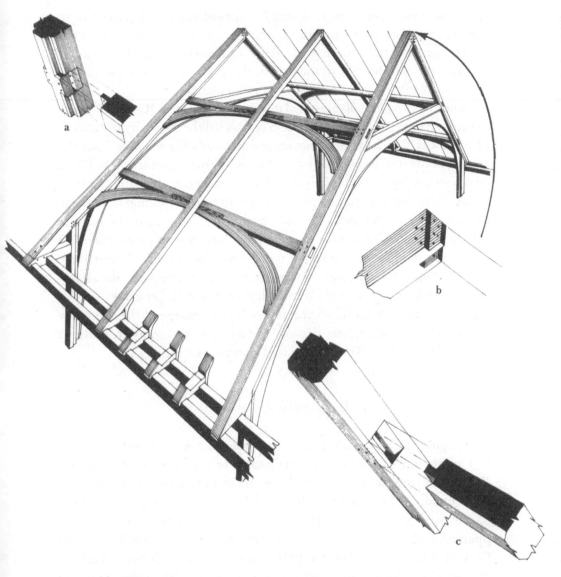

57. The roof of King's College Chapel, Cambridge, showing the different side-purlin joints used *c*.1480 (above) and 1510-12 (below).

two side-purlins in each slope and wind-bracing of the Tudor style — all of which is attributable to Master Martin Prentice who, in about April 1480 bought a skin of parchment on which to draw his design for it. Five bays of his roof were built and clad between 1476 and 1485, when the western end was temporarily blocked and the church put into use. The patination of exposure to the weather can still be seen on the fifth frame from the eastern gable, deriving from this period. The joints securing the side-purlins to the principal-rafters are as shown in inset *a*, single and central tenons without any refinements.

When work eventually resumed it was in the charge of Richard Russell, master carpenter to Henry VIII, who completed the work. By the middle of August 1512 he had the trusses framed and ready for erection. Then the rearing of the roof began, and by the middle of that September it had been clad with lead by the plumbers (Dr. J. Saltmarsh, personal communication, 1973). Thereafter the same carpenters began to raise scaffolding and centres for the vaulting operation. The side-purlins were jointed in the manner shown in inset *c*. This was the most efficient form of tenon ever evolved to resist shear stress, against which it is effectually buttressed (Hewett 1980, 283-7). It provides a precision dating instrument that prevents the ascription of any buildings with such haunched tenons to earlier dates than this joint's inception. Russell had previously been in charge of the carpentry at Westminster Abbey, later that of St Margaret's, Westminster, Wolsey's works at York Place, Westminster, and Hampton Court.

Bath Abbey, the Nave

The high-roof of the nave of Bath Abbey, a bay of which is shown in Fig. 58, must, it seems, be attributable to the brothers Robert and William Vertue, who were the designers of the masonry, the dates of their buildings being from 1501 until 1539. It is a low-pitched single hammer-beamed design without any particular merit, intended as an open roof, since the existing vaults were introduced in 1864-73 by Sir George Gilbert Scott (Harvey 1974, 34, 195, 218). As shown at right in the drawing, the common-rafters were fitted into the ridge-piece by means of tenons with diminished haunches.

Abbey Dore, Herefordshire

The Abbey was dissolved in 1535, and the buildings were allowed to fall into decay. John Abel, 'architector and carpenter', restored them

58. Bath Abbey, the roof of the nave.

for the first Viscount Scudamore in 1633-4. He built a new roof with attached ceiling of local oak, with straight tie-beams spanning 32 feet (Colvin 1948, 235-7). The bays contain 10 common-rafters each, with double wall-plates. Four purlins were fitted each side, and a ridge-piece, the principal rafters supported by three tiers of vertical struts. Fig. 59 illustrates the structure, with details of the carved pendants below the tie-beam and the scrolled corbels on the wall-posts, which are mounted on the stone columns of the original building.

Wells Cathedral, the North Transept

The latest of the roofs of Wells is that above the vaults of the north transept, illustrated in Fig. 60. This work would appear to date from 1661,

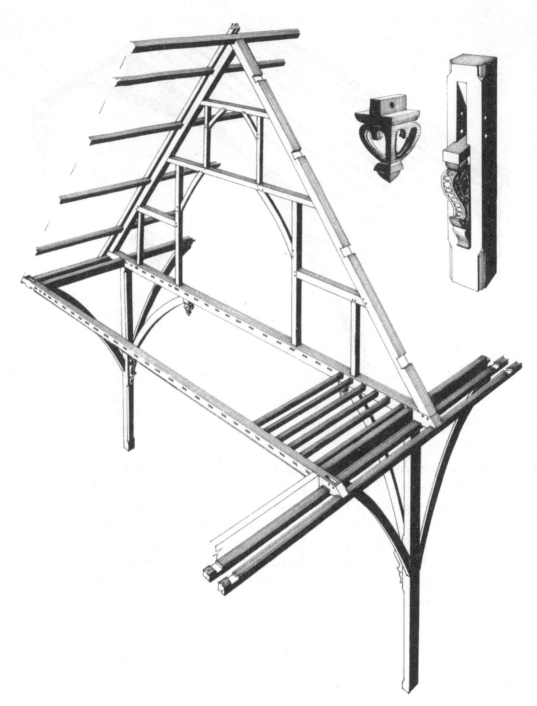

59. Abbey Dore, Herefordshire, the roof of the nave.

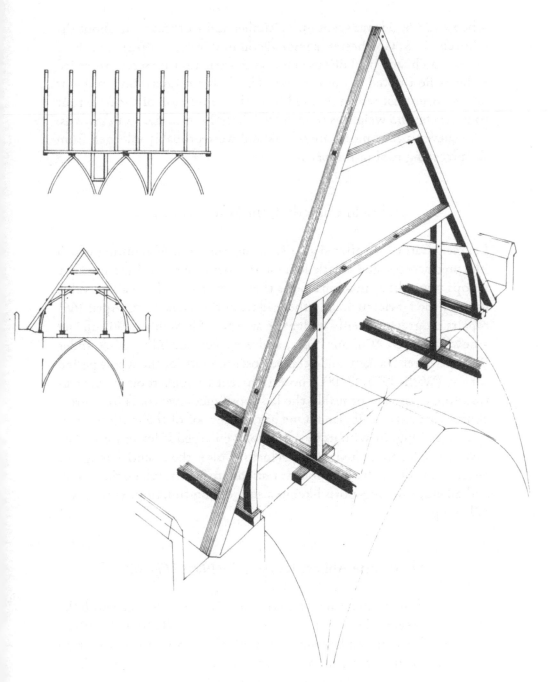

60. Wells Cathedral, the high-roof of the north transept.

when £227 6s 3d was spent upon 'timber and workmanship about the Church' (L. S. Colchester, personal communication, 1972). This is of pine or larch, and the bolts used for its construction are screw-threaded – the earliest noted in this context. The design is interesting in that it derives from that which resulted from the conversion of the nave roof to parapets; the weight is transmitted directly onto the vault crowns. The curved ashlar-pieces are of oak and were probably salvaged from the preceding roof of this part.

Lichfield Cathedral, the South Transept

Lichfield Cathedral suffered greatly during the Civil War, during which 'the precincts became a battlefield' on two occasions, and Parliamentary troops wrought as much havoc as they were able, selling all movable parts of the fabric, including the lead from the roofs. From April 1661 repairs progressed evenly under the architect Sir William Wilson and a new Dean, Dr William Paule, and were completed by 1669 under Bishop Walter Hackett, and detailed cost accounts for the whole period survive (Wallis 1970, 17-18). Those 17th-century high-roofs are in position today, and are probably the best post-medieval roofs for a great church that exist in England. One bay of the roof of the south transept is shown in Fig. 61 with its tie-beam joint enlarged at lower right. The eaves triangles are bolted with forelocks. This is the essential design of all the roofs of the Restoration; variations are subtle, where they exist, and all employ king-posts having a supporting function, with dove-tailed tops.

Sherborne Abbey, Dorset, the Nave (*Fig. 62*)

This part of the church dates from the late 15th century when both the choir aisles were rebuilt in the Perpendicular style (Gibb 1972, 30-1). The roof above the vaults cannot be original, but appears to date from the 17th century; it is a simple queen-posted design well wrought in oak, without recourse to any form of ironwork.

61. Lichfield Cathedral, the high-roof of the south transept.

62. Sherborne Abbey, Dorset, a roof frame from the nave.

Tewkesbury Abbey, the Nave

Tewkesbury Abbey church was re-roofed above the nave vaults at an unknown date, and the frame design seems to predate that of the nave of St Paul's by Wren. A single truss of this is shown as Fig. 63. It was framed in well-tooled oak with the least possible recourse to iron fastenings. This is a king-post roof, the posts being in extension, with raking-struts and common-purlins. The purlins proper, which are the two illustrated, were fitted into the principals by diminished-haunch tenons. The feet of the king-posts were jointed to the tie-beams in a remarkable manner, being dovetailed through the beams, and their

63. Tewkesbury Abbey, the high-roof of the nave.

taper compensated by driving wedges downward beside them. One forelock bolt was used for each.

St Paul's Cathedral, the Nave

The nave roof of St Paul's, shown in Fig. 64, was designed by Sir Christopher Wren and built between 1696 and 1706. It is a king-post roof with pendant posts secured to the tie-beams with iron stirrups and screw-threaded bolts. The jointing for the feet of the rafters, the ends of the tie-beams and the wall-plate scarfs are shown as insets. The span is a little over 50 ft., and Wren had difficulty in finding oaks long enough for the tie-beams. It has been stated that the quest for these 50 great trees took several years; they were eventually offered by the Duke of Newcastle in 1693, but did not arrive on site until 1696 (Lang 1956, 175-6 and 228).

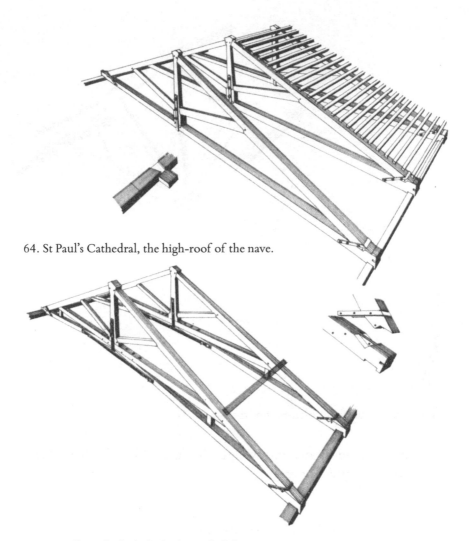

64. St Paul's Cathedral, the high-roof of the nave.

65. St Paul's Cathedral, the high-roof of the west portico.

St Paul's Cathedral, the West Portico

Wren's roof over the west portico of St Paul's is illustrated by one frame in Fig. 65. This is a variation of his roof for the nave, and it is surprising that he did not avoid his long search for 50-ft. oaks, when he might more intelligently have used this design throughout the length of his cathedral. This is the seemingly inevitable king-post with raking-strut design, from which Renaissance architects never fully escaped, but the whole is mounted upon the collars which with the other components form built camber-beams.

Durham Cathedral, the Choir

The earliest high-roof surviving at Durham Cathedral is that above the choir, of which one bay is shown in Fig. 66. It probably relates to the restoration by the architect James Clement who died in 1690 (The Ven. C. J. Stranks, personal communication, 1973). It is of oak and

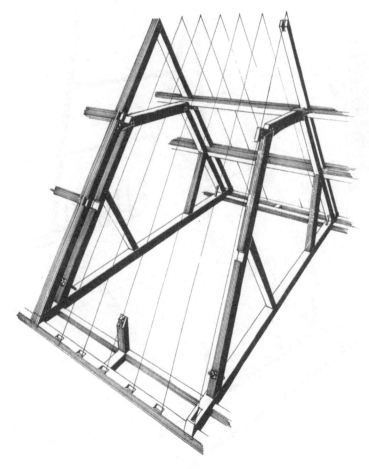

66. Durham Cathedral, the high-roof of the choir.

has under-rafters supporting its collars, which are lap-dovetailed and in turn support the upper side-purlins. All the purlins are doubled, as shown, and the eaves triangles are secured with screw-threaded bolts – which may be even earlier examples than those used in the northern transept of Wells (Fig. 60).

Winchester Cathedral, the east end of the Nave (*Fig. 67*)

Several bays at the east end of the nave at Winchester were re-roofed, and I am assured that documentary evidence exists for the date of this

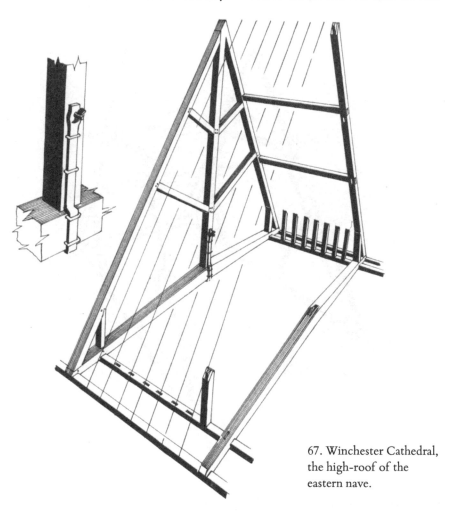

67. Winchester Cathedral, the high-roof of the eastern nave.

work: 1699 (W. J. Carpenter-Turner, personal communication, 1973). Five king-post trusses exist, defining bays containing seven common-rafters each, and having ashlar-pieced feet. There are two side-purlins to each slope, and two raking-struts — the foot stirrup-irons of the posts being forelock-bolted and stapled as shown at left.

Worcester Cathedral, the Eastern Roofs

The three eastern arms of Worcester are roofed as illustrated in Fig. 68. These are king-post and raking-strut trusses wrought in oak, with some

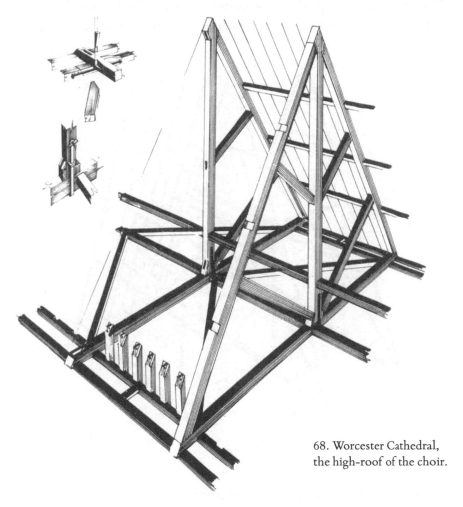

68. Worcester Cathedral, the high-roof of the choir.

ironwork that is everywhere forelock-bolted. An unusual feature is the forcible bracing of the horizontal plan shapes of the bays by diagonal timbers wedged into bird-mouthed sockets. In addition, the king-posts were steadied at a height of roughly three feet above the tie-beams by purlins tenoned into them – indicating an erection method of one bay at a time. Ashlar-pieces were fitted with chase-tenoned tops, and lapped and housed feet, while the cladding was carried by purlins and common-rafters.

Lincoln Cathedral, the Great South Transept (*Fig. 69*)

This roof is of pine, probably of Baltic origin, and its date would appear to be after the opening of the 18th century – subject to documentation.

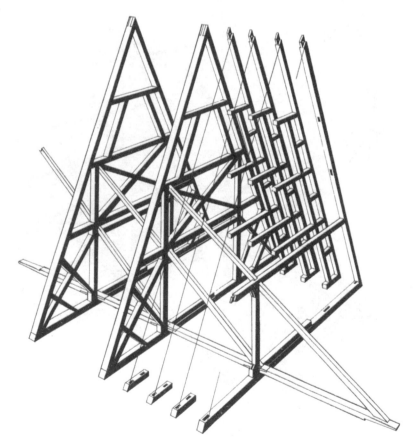

69. Lincoln Cathedral, the high-roof of the great south transept.

Various numbering systems are visible but they seem to relate to the timber as a quayside cargo rather than as a prefabricated structure – as noted also at 'G' Warehouse, St Katharine's Dock, London (Hewett 1980, 249-252). It comprises ten trusses with three common-couples between them. The lesser couples appear to belong to an earlier system, stiffened by under-rafters to their third collars. The first destruction of the roof must have been associated with the fall of the central spire in 1549, but the present roof is unlikely to date from then. The rack-resistance measures built into this roof are considered to be later than the roof, and to relate to the purlins laid over the tie-beams during the 18th century.

Lincoln Cathedral, the Great North Transept (*Fig. 70*)

This is, again, of pine, comprising nine trusses having common-purlins. The northern truss of this roof is a bolt-tensioned device resembling those used by Wren at St Paul's, to which there is a similar counterpart in the gable of the south transept; dates after *c.*1700 are proposed. The vertical bolt appears to be tightened by a forelock.

Beverley Minster, the Great South Transept (*Fig. 71*)

This roof may be part of the work undertaken by Nicholas Hawksmoor in the period 1715-36. It is of oak, some of it re-used, with four common-rafters between tie-beams. The tie-beams are doubled, placed upon one wall-plate and under the other, with four threaded through-bolts. The principal-rafters are double also, with three side-purlins and a ridge-piece. In each truss there is a king-post with two braces, and two queen-posts each with two braces. It is the best roof of the later designs.

Ely Cathedral, the Choir (*Fig. 72*)

The high-roof above the vaults of Ely's choir is by James Essex, and several of the timbers are dated May 1768, which is after its restoration. It

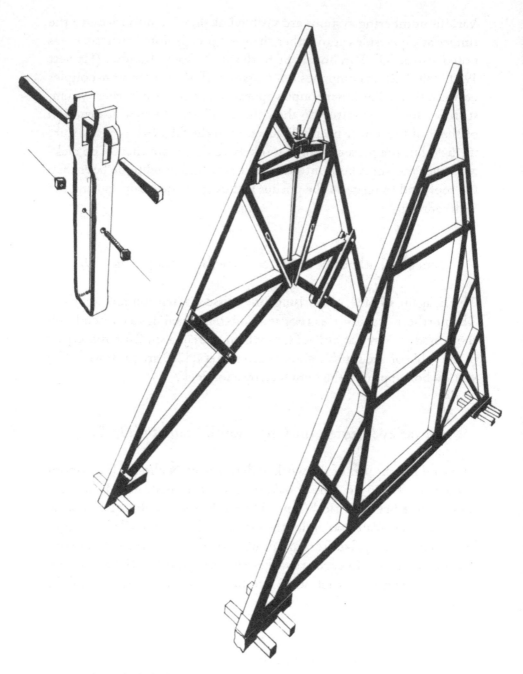

70. Lincoln Cathedral, the high-roof of the great north transept.

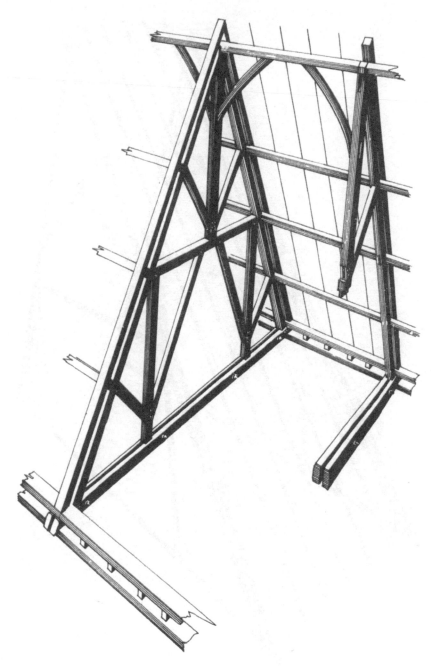

71. Beverley Minster, the high-roof of the great south transept.

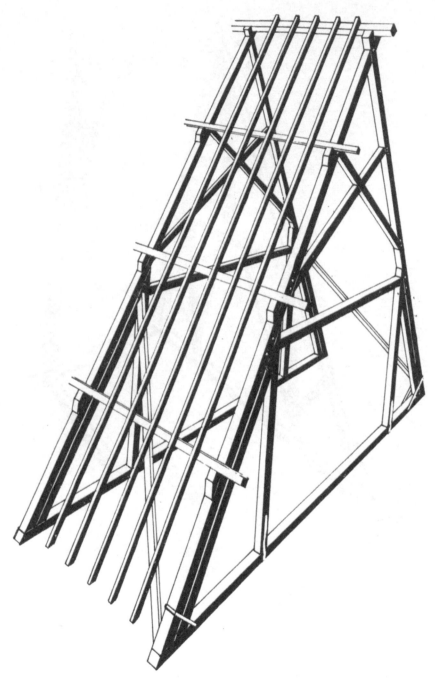

72. Ely Cathedral, the high-roof of the choir.

is of pine, most of which is side-axe dressed, and the design is of queen-posts with scissors above the collars. The pitches were braced against racking, and the ironwork fastened with forelock-bolts.

York Minster, the South Main Transept

Between *c*.1770 and 1780, a new timber roof was built to cover the medieval timber 'vault' of the south main transept of York Minster by L. Tebb (Dr. E. A. Gee, verbal information, 1972). This roof had to clear the ridge-rib of the 'vault'; it did so without recourse to stilted frames, as shown in Fig. 73. This is a collar-king-post roof having eight cants

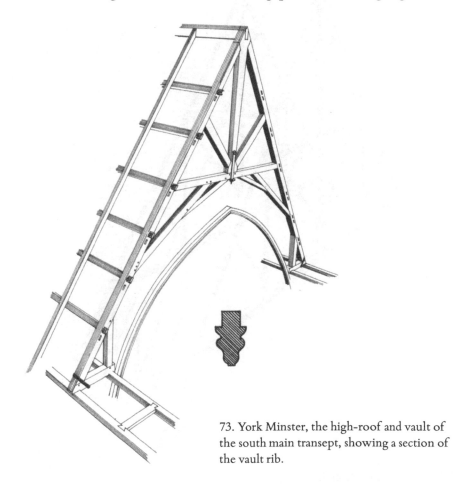

73. York Minster, the high-roof and vault of the south main transept, showing a section of the vault rib.

framed beneath its collars; the design draws much upon medieval and Renaissance traditions. Both common side-purlins and common-rafters were used.

Canterbury Cathedral, the South-East Transept

The south-east transept has an impressive roof, of the high type, surmounting the vaults, illustrated in Fig. 74. It has the date 1771 carved on a tie-beam, and is framed in oak, creosoted. Good trusses combining

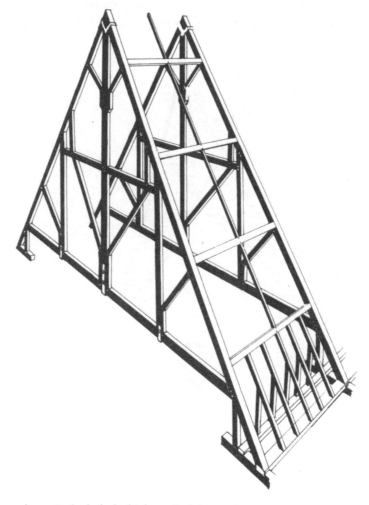

74. Canterbury Cathedral, the high-roof of the south-east transept.

both king- and queen-posts are set stilted, their ties being above the ashlar-pieces.

Rochester Cathedral, the Great North Transept (*Fig. 75*)

When the great north transept of Rochester Cathedral was considered to need a new roof in 1825, the architect L. N. Cottingham did the work, and at this time the roofs were restored to the medieval pitches indicated

75. Rochester Cathedral, the high-roof of the great north transept.

by the torchings on the crossing tower. This is almost entirely built of softwood and is unusually intricate. It would appear that Cottingham knew the inherent weakness of pine, and sought to avert failures by this design. The complex beneath the collars forms a sagging-truss for the built tie-beams, and above this a suspended king-post and raking-strut truss is mounted. The eaves assembly is shown at right.

York Minster, the Choir

The present high roof over the nave vaults of York Minster is by Sir Robert Smirke and is datable to the restoration of 1829-32. The iron castings are by Messrs. Harwood & Dalk of York, and are dated 1829.

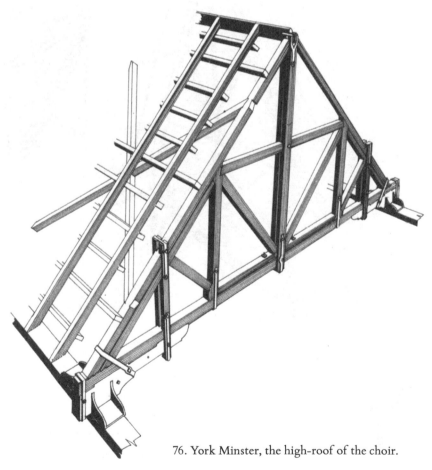

76. York Minster, the high-roof of the choir.

The span being a little over 50 ft., the truss design had to be elaborate, as may be seen in the illustration, Fig. 76. This is essentially a queen-post design with raking-struts and collars, to which were added clasping king-posts and clasping-struts – the whole being well provided with iron strapwork and bolts. Each pitch-plane was saltire-braced, and common-purlins support the common-rafters. Both oak and pine were used, the tie-beams being of the former.

York Minster, the Nave

The roof of the nave is by Sydney Smirke, occasioned by the restoration after the fire of 1840. This is illustrated in Fig. 77. It is of mixed oak and

77. York Minster, the high-roof of the nave.

pine, and all bolts are screw-threaded. It has twelve common purlins in each slope; an interesting development in this case is the setting of the lowest purlins directly on top of the ashlar pieces.

Sherborne Abbey, Dorset, the Choir

The choir of Sherborne Abbey has a very early and boldly-designed fan vault, which had begun to subside seriously by 1856. The whole fan vault was rebuilt, the operation being completed in three months at a cost of £425 (Gibb 1972, 19). It seems probable that this may also be the date of the high-roof, a part of which is illustrated in Fig. 78.

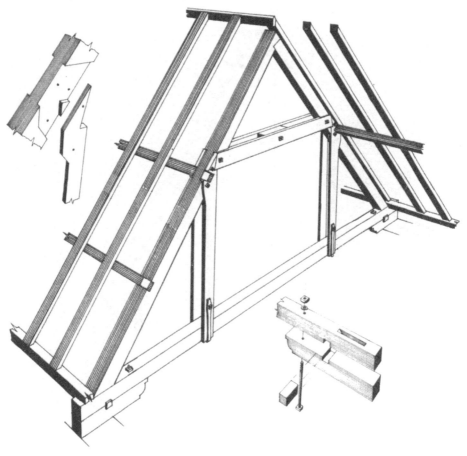

78. Sherborne Abbey, Dorset, the high-roof of the choir.

Much use was made of screw-threaded nuts and bolts. The queen posts were paired, as were the collars, and fitted into remarkably elaborate housings on the principal-rafter flanks. The tie-beams were 'built' at the eaves, and an example of this is shown to the lower right of the drawing, but no explanation can be offered for this feature, except that the architect was evidently concerned about inflexibility at that point.

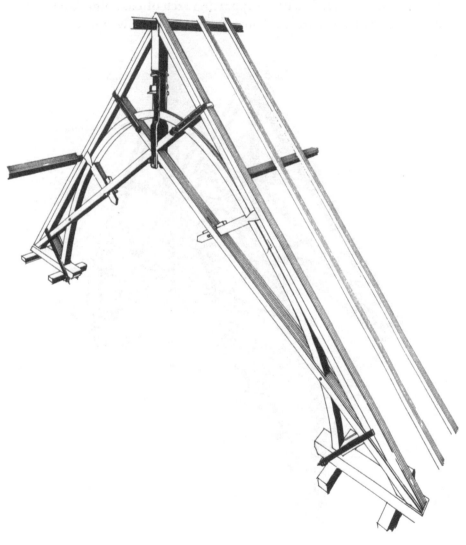

79. Rochester Cathedral, the high-roof of the choir.

Rochester Cathedral, the Choir

The choir of Rochester has the high roof designed by L. N. Cottingham in 1825, illustrated by Fig. 79. This is a design that compels some admiration, wrought in mixed oak and softwood. In it are combined many elements that had previously been found adequate by themselves, and it would be of interest to know how long it might endure. It is scissor-braced, under-raftered, with a supporting arch of oak, high king-posts and raking-struts that are mechanically tensioned by wedging.

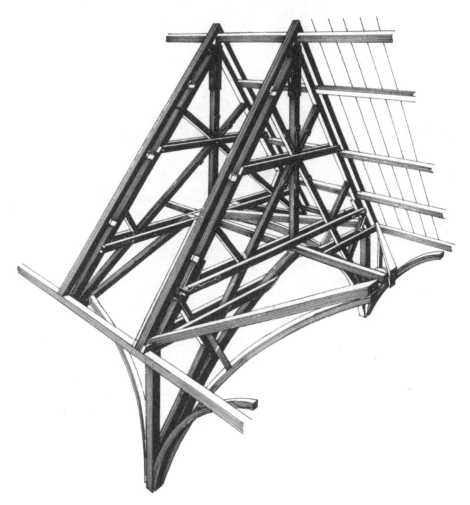

80. Ripon Cathedral, the high-roof of the choir.

Ripon Cathedral, the Choir

The existing high roof above the choir of Ripon Cathedral may date from the restoration by Sir George Gilbert Scott, carried out between 1861 and 1869. The ironwork used in its design is galvanised, and although the process was invented in 1742 this work appears later, by comparison with any dated 18th-century works. Two trusses of this are illustrated in Fig. 80, one with wall-posts for vault pockets and the other to clear their crowns. The whole was designed with an integral wooden 'vault' beneath it, and employs scissor-braces with intersecting king-posts, both clasped by dual collars and reinforced by six-armed stars of galvanised iron. Like the roofs by Cottingham this will no doubt interest posterity, since it might show that good design can extend the life of poor materials – but this is most improbable.

LEAN-TO ROOFS

Wells Cathedral, the Nave Triforia

The lean-to roofs of the nave triforia at Wells are probably the oldest that survive in England. One principal-rafter from the northern range is shown as Fig. 81. This is probably of the period 1175–*c*.1192 (Harvey 1974, 241), and its design relates well to that of the oldest main-span frames over the nave. It has single collars supporting side-purlins – timbers which, in the lean-to situation, cannot have been used to resist racking tendencies because there are none; their early introduction into such roofs can only have been to preserve flatness of the pitch-plane. As in the roof of the choir, these lean-tos were converted from eaves to parapets before *c*.1320, and the base triangulation and wall-plate systems were lost at that time.

Salisbury Cathedral, the North Choir Triforium

The finest lean-to roof surviving in any English cathedral is that illustrated in Fig. 82, above the northern triforium at Salisbury, beside the choir; this part was built between 1225 and 1237, the architect being Nicholas of Ely (Harvey 1972, 157).

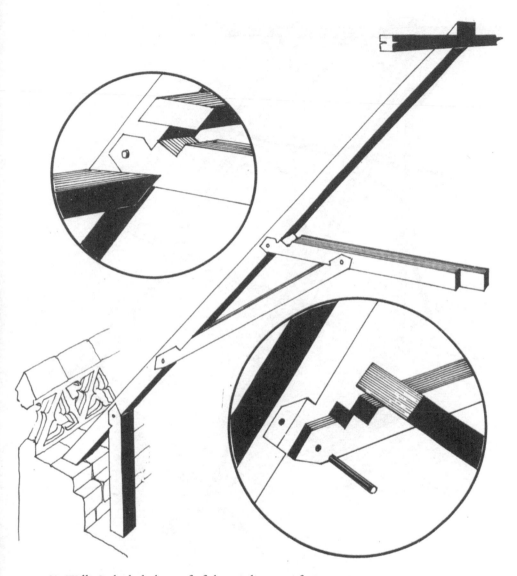

81. Wells Cathedral, the roof of the north nave triforium.

The drawing shows two closely-spaced frames of the roof, which is an elaborate essay in framing, and an example of anti-racking consider-ations being applied to the lean-to situation. In this case it was justified, because the roof was designed as an irregular-sectioned timber building

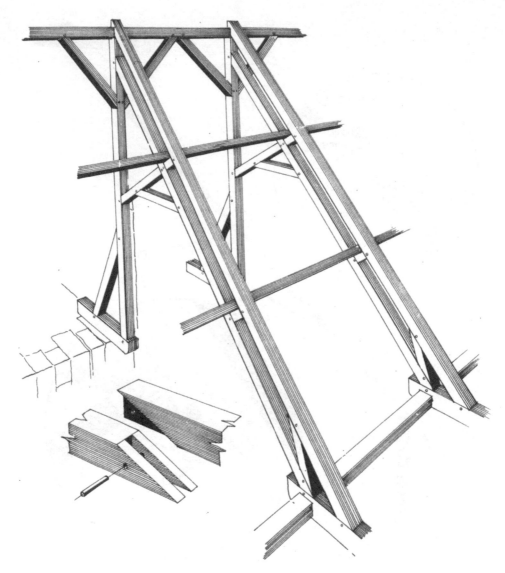

82. Salisbury Cathedral, the roof of the north choir triforium.

that could very well stand alone and independently of its situation on the masonry carcase. The posts have triangulated feet and the pitch-plane is under-raftered, whilst raking-struts convert sagging tendencies down to the sole-pieces. Two purlins occupy each slope and the top-plate is set into trenched stone corbels, or plate-hooks, and framed as an 'arcade' that is inherently stable. The scarf-joint used here is through-splayed and bladed.

Lincoln Cathedral, the North Triforium of the Choir

The north triforium of the choir at Lincoln retains the graceful roof illustrated in Fig. 83, in which a restrained curvature was introduced. Harvey's date, 1192-1210, is assumed to relate to the sub-structure (Harvey 1974, 230). The whole was un-tied, probably in view of its deliberately lightweight construction, and the masonry was expected to contain all of its outward moment. No notched lap joints were used for this, but the splayed and tabled scarf is indicative of a 13th-century date.

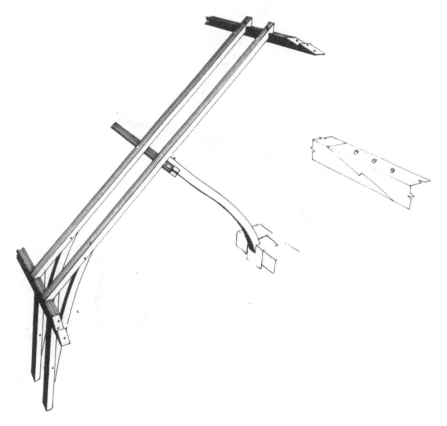

83. Lincoln Cathedral, the roof of the north choir triforium.

York Minster, the Triforium of the North Transept

Fig. 84 shows a representative raking-strut of the branching type which exists in the triforium of the north transept of York Minster. The fabric beneath is dated to *c*.1230-1241 (Harvey 1972, 169) and the timber roof to between *c*.1361 and 1373 (Dr. E. A. Gee, verbal information, 1972). All the rafters are common and of uniform cross-section, and the designer's main concern was the stability of the pitch-plane and its flatness. The work was, like that at Lincoln, dependent upon its masonry situation. The introduction and development of jowls at the ends of the

84. York Minster, the roof of the north transept triforium.

raking-struts is reflected here – a technique developed soon after *c.*1250 (Hewett 1980, 101).

Rochester Cathedral, the South Choir Aisle

The south choir aisle of Rochester Cathedral came, in St John Hope's view, into its present state shortly after 1322, and the low-pitched and 'vaulted' timber roof of it are clearly of that date (Hope 1900, 71-2). This is illustrated in Fig. 85, in which the profile of the ribs is shown enlarged and the panelling omitted. This roof was designed to stand on its carved timber corbels, and its posts to support the low-pitched roof-plane, into which the truncated tunnel-ceiling was framed. The angled tying timbers, fitted to the ribs at mid-arc, are a feature frequently noted in 14th-century contexts, and the use of jowls for the posts confirms the

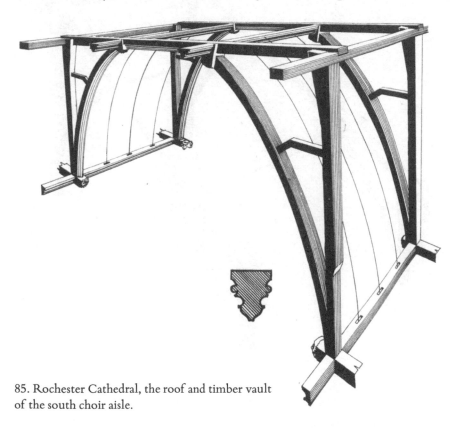

85. Rochester Cathedral, the roof and timber vault of the south choir aisle.

dating (Hewett 1980, 101). Traces of the original colouring of this roof were discerned on the rib-timbers – red for the rounds, green for the hollows, and white for the fillets.

Wells Cathedral, the North Choir Aisle (*Fig. 86*)

The north choir aisle of Wells was completed in either 1335 or 1380 (L. S. Colchester, verbal information, 1972), but whichever is proved will date its roof – the original one. The posts stand directly on the masonry, either at vault-crown or in the pockets, and are jowled. The

86. Wells Cathedral, the roof of the north choir aisle.

low-pitched rafters are supported by curved braces similar to those at Rochester, and the outer purlin was trapped by the little tie-piece.

Cartmel Priory, the South Choir Aisle

The church of Cartmel Priory was founded on its present site in 1190, during the Great Transition (from Romanesque to Early English) by William Marshall, Earl of Pembroke. Eventually the south choir aisle, known as the Town Choir, was built in *c.*1340 (Dickinson 1978, 10-12), and this retains its original lean-to roof – illustrated in Fig. 87. This

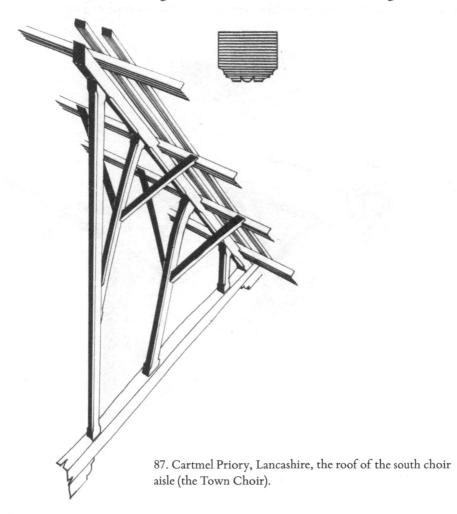

87. Cartmel Priory, Lancashire, the roof of the south choir aisle (the Town Choir).

appears to be an early example of the type having vertical and raking-struts, from which braces spring to support the purlins – of which there are three. All timbers are nicely stop-chamfered and the tie-beam is moulded as illustrated.

York Minster, the Nave Aisles

The roofs of the nave aisles at York are unlike others of their kind; an example is illustrated as Fig. 88. The nave itself was built between 1291 and 1345 under Master Simon Mason (Harvey 1974, 246), and the roofs to the aisles were probably begun *c*.1361 and completed by *c*.1375

88. York Minster, the roof of the nave aisle.

(Dr. E. A. Gee, verbal information, 1972). These frames are remarkable, being low-pitched and tie-beamed, but designed as inflexible triangles – the posts jowled and in the practicable cases sole-braced. The sole-pieces for the common-rafters are unusual and are designed to hook over the stone top course, and to resist rafter-slide outwards. The scarf-joints used are stop-splayed and prolifically pegged.

Carlisle Cathedral, the North Aisle Triforium

The triforium above the north aisle of Carlisle Cathedral choir is the original and dates from between 1363 and 1395, when the upper choir walls were built, possibly under John Lewyn (Harvey 1974, 221). This low-pitched lean-to roof is illustrated by Fig. 89, which shows a

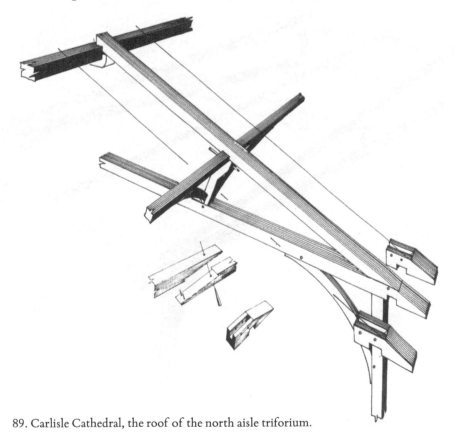

89. Carlisle Cathedral, the roof of the north aisle triforium.

single tie-beam, wall-piece, arch-brace, jowled raking-strut, purlin and top-plate – the last set into a trenched stone corbel or plate-hook. The wall-plate was single, and passed through trenches beneath the sole-pieces – one of which is shown enlarged. The scarf used is stop-splayed with square under-squinted butts – and accords precisely with the fabric dates, having both iron spikes and a single face-peg.

Canterbury Cathedral, the Nave Aisles

The low-pitched roofs of the nave aisles at Canterbury are framed throughout as illustrated in Fig. 90, with 15 frames to both north and south series. The timbers are of massive sections and have survived in

90. Canterbury Cathedral, the roof of the nave aisle.

good order – for that reason. The curved braces from the wall-posts to the transoms originally existed at both inner and outer ends, and the sections of the wall-posts are double-hollowed with an intervening return – a moulding most used between *c.*1370 and *c.*1550 (Forrester 1972, 31). The details of the assembly are as illustrated, with five common-rafters in each bay that are let into the top-face of the central purlin and butt-cogged into the outer wall-plates, which tenon into the transoms' flanks. The known dates for the building of the nave, under Henry Yevele, are from 1379 until 1405, and no features of the roofs described conflict with those dates (Harvey 1974, 221).

Norwich Cathedral, the North Triforium of the Nave

The brilliant essay in low-pitched lean-to roofing illustrated in Fig. 91 formerly existed over the northern nave triforium of Norwich

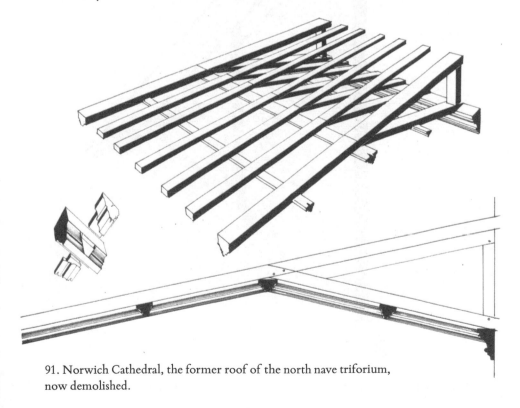

91. Norwich Cathedral, the former roof of the north nave triforium, now demolished.

Cathedral; it has recently been replaced, necessarily by something inferior. The scope for invention and the unpredictability of genius were both well displayed in this work, which was second – if at all – only to the lean-to roof of Salisbury's triforium. The drawing is as

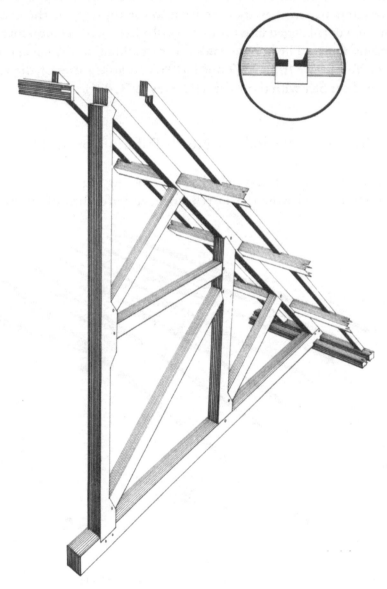

92. Malmesbury Abbey, Wiltshire, the roof of the south triforium.

deceptive as the roof was, since it appears to be a lean-to roof, as did the original — but that was not the case, for it was constructed as a low-pitched ridged roof and simultaneously converted into a single slope. Every rafter, whether principal or common, was built from three components of which the first two formed a ridge, and the third converted it to a pitched-plane. No width tying was provided, and its survival always depended upon the two imposts of masonry and the efficiency of its triplicated central joint — which was not examined when taken apart. The profiles of its mouldings had a wide period of usage but their popularity covered the years between *c*.1370 and *c*.1550 (Forrester 1972, 31). No tenon haunches or soffit spurs were used.

Malmesbury Abbey, Wiltshire, the South Triforium

The south triforium of Malmesbury Abbey church retains the lean-to roof shown in Fig. 92. The pitch was inherited, and steep, and the carpentry is direct, unimaginative and self-reliant — being of good quality oak and having no recourse to iron reinforcements. The three purlins were fitted by tenons having diminished haunches, as shown inset.

St Paul's Cathedral, the Triforia

With Wren's work at the new St Paul's a firm date is again available, 1691-2, (Lang 1956, 143-6) and the standard of carpentry is surprisingly high for that building, where great expense was evidently incurred to provide well-moulded timber in all such situations as were to be subsequently visible. One frame of a triforium roof is shown in Fig. 93. This was a design that stood on points well below the level of the low-pitched roof-plane, supported on tall posts that were planed into the section having 'rounds' at each corner, shown at *a*, while the tie-beams were planed into cyma rectas at their lowest arrises shown at *b*. All the oak timbers are very heavy for their functions and, although they are perfectly adequately jointed and pegged, in most cases they are reinforced by means of iron bolts with threaded nuts.

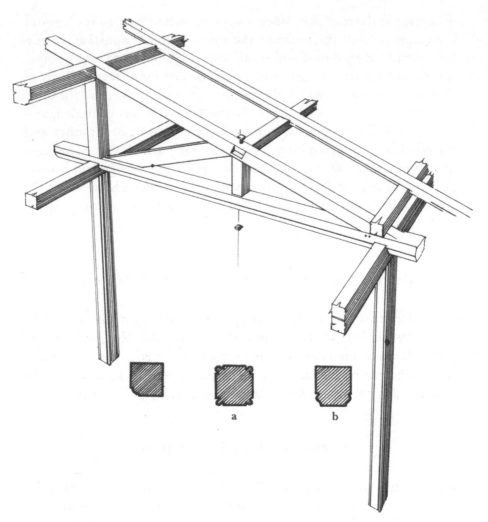

93. St Paul's Cathedral, a triforium roof.

Lincoln Cathedral, the North Triforium of the Nave

The northern nave triforium at Lincoln is roofed mainly in pine with trusses as illustrated in Fig. 94. No date for this is available, except that of the restoration by James Essex between 1762 and 1765. Some of the rafters are of oak; they incorporate stop-splayed scarfs and for that reason are likely to be the spoils of the previous roof in this situation. The use

94. Lincoln Cathedral, the roof of the north nave triforium.

of compassed butment-cheeks for the two canted struts is of interest; it shows the aim to distribute the stress equally between the two, when the foot-bolt, with a pocketed screw-threaded nut, was tightened prior to setting the purlins on. This type of foot-bolt with pocket-nut was not used by Wren, so far as is known, but exists in the present roof of Henry VII's Chapel, Westminster Abbey, inscribed 1785. The post-head cut to hold the purlin is shown, as is also the scarf used for that purlin.

Salisbury Cathedral, the South Triforium of the Nave

The lean-to roof illustrated by Fig. 95 is over the south triforium of the nave at Salisbury. It varies from what was evidently the general 18th-century practice, in that its main raking-strut passes through the post, which is chase-tenoned into notched butment-cheeks; the short lower part is merely nailed to its soffit. Overall these frames were well jointed, and they supported common-purlins. I am indebted to Mr. R. O. C. Spring, of the cathedral, for detailed information concerning this roof.

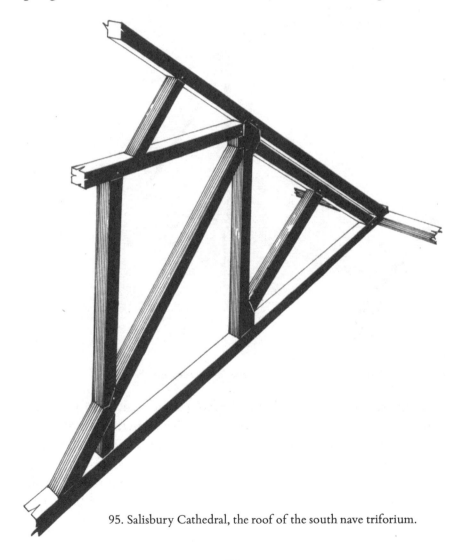

95. Salisbury Cathedral, the roof of the south nave triforium.

1. The spire of Salisbury Cathedral. Painting by the author.

2. Rayleigh Castle, Essex. A post excavated by Mr. E. B. Francis in 1910. (*Photograph: Peter Richards*)

3. A notched lap joint of archaic profile which has broken under extending stress. From the belfry of Navestock church, Essex.

4. A secret notched lap joint without refined profile, from Romsey Abbey, Hampshire.

5. Portrait of Adam Lock, master mason (d. 1229), from Wells Cathedral.

6. An open notched lap joint of refined profile from Wells Cathedral. It was cut with a side-axe.

7. Woodcut by Hans Schäuffelein, *c*.1515, from the *Weisskunig*. This shows two types of axes which were in use at the time. A broad-axe is being used to dress the timber to shape, along the grain.

8. From the Barley Barn, Cressing Temple, Essex. This shows the use of a square, on which the mortise was marked out with a sharp knife. There is no indication of what kind of tool was used for the vertical cutting of the mortise. (*Photograph: John McCann*)

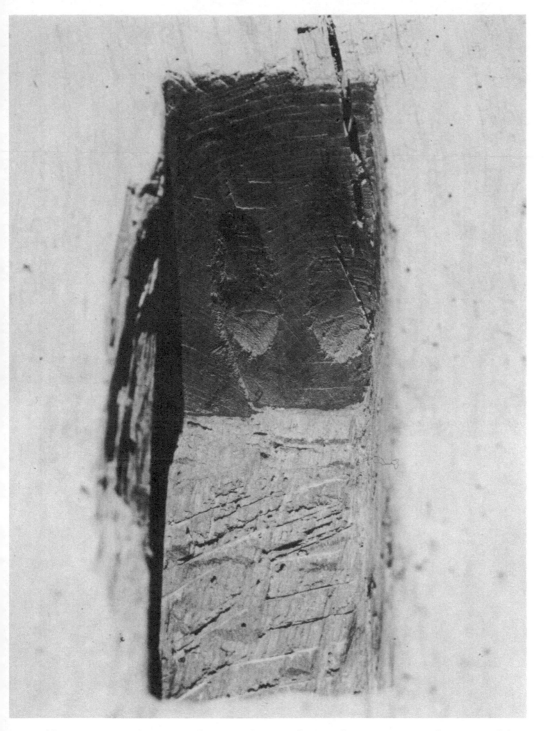

9. An oblique mortise in the same Barley Barn, showing the use of a spoon auger to clear some of the waste before cutting the joint to shape. It also shows the marks of a 2½-inch chisel which was used with a large, heavy mallet. (*Photograph: John McCann*)

10. Wells Cathedral, the timber roof over the Chapter House; probably the finest in England. (*Photograph: Caroline Malone*)

York Minster, the West Triforium of the South Transept

The lean-to roof shown in Fig. 96, is of the west triforium of the south transept of York Minster. It was designed to resist racking movement, which cannot normally occur in a lean-to situation. The purlins are tusk-tenoned and the bolts holding the braces to the raking-struts are screw-threaded.

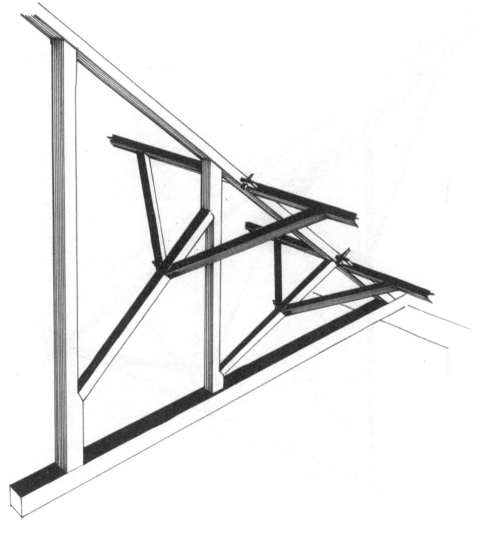

96. York Minster, the west triforium roof of the south transept.

Ely Cathedral, the Triforium of the North Transept

The north transept triforium at Ely has the lean-to roof shown in Fig. 97. This was a softwood construction, and somewhat better than most, since all timbers were properly jointed, without recourse to iron bolts; only the eaves angle of the under-rafter was strapped. The purlins were tenoned into the posts.

97. Ely Cathedral, the triforium roof of the north transept.

ROOFS OF POLYGONAL PLAN

Westminster Abbey, the Eastern Apse

Only four apsidal high-roofs survive in this country, and of those but one is medieval. This is a splendid specimen, however, and roofs the eastern arm of Westminster Abbey, which is illustrated in Fig. 98 as though viewed from within, and showing structural components postulated by original matrices, empty at the time of examination. The whole has been dismantled and re-erected during recent years but should still sustain serious study, as the sole survivor of a former tradition for polygonal high-roofs at the highest possible level – the great or cathedral church. The three eastern arms of Westminster Abbey were completed by 1259 under the direction of Master Alexander, Henry III's Master Carpenter (MacDowall, Smith and Stell 1966, 156), and the eastern arm may be presumed to have been finished first. The plan to be roofed in this case was five-sided, demi-decagonal; the two tie-beams shown were set across this figure and the two remaining angles tied by diagonal timbers radiating from the eastern beam. The couples, comprising 28, were fitted each with two collars and scissor-braces, doubled wall-plates, and ashlar-pieces, as illustrated. The feet of the scissors were secured by secret and diminished notched lap joints, as shown

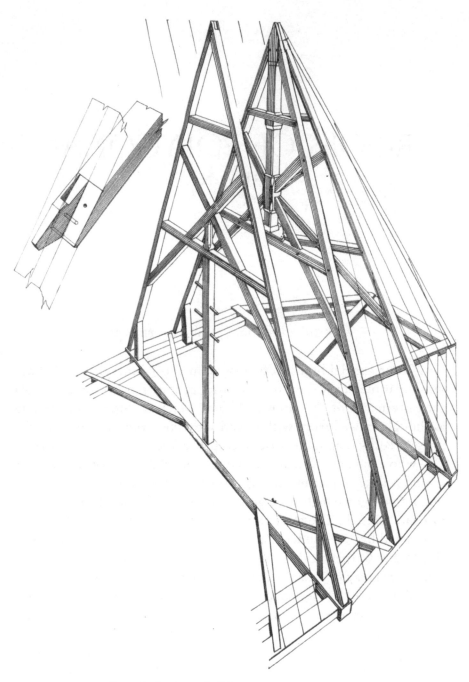

98. Westminster Abbey, the high-roof of the eastern apse.

enlarged. At the eastern couple a king-post was mounted, which stands on a large plate of oak decoratively shaped and pegged to the several collars beneath it. The 'halfscissors' within the polygonal half-conoid termination chase-tenon into the post, which was left suitably thick at that point. The tie-beams were fitted with branching ends, as had been done previously at Salisbury, and the ashlars originally stood upon the inner of the two plates, which was tenoned into the flanks of the tie-beam. The common-rafters in the facets run parallel, and were tapered off to meet the flanks of the principals, into which they fade, and were spiked. The crown-post standing centrally on the tie-beam rises to the height of the crossing of the scissor-braces.

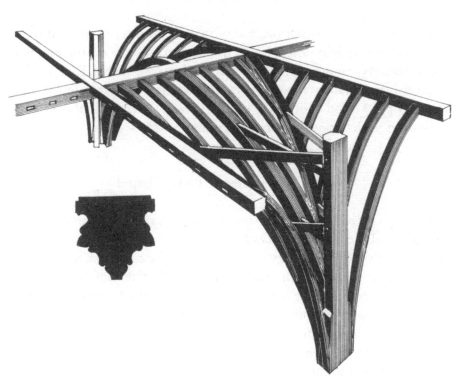

99. St Albans Cathedral, the timber vault of the presbytery.

St Albans Cathedral, the Timber Vault of the Presbytery

The timber vault of the presbytery is the oldest of the kind in England, and is shown in Fig. 99, with the profile of its rib-timbers. This is diagonally supported from posts that formerly may have been components of the high roof, now missing. 'The remains of vaulting shafts internally and of abutments for flying buttresses on the exterior are clear proof that wooden vaults were never common in the greater churches and, since the fires at Selby and York, that at St Albans is probably the earliest of any importance still in existence. ... At the north end of the east wall of the presbytery is a painting of a figure in Archbishop's vestments and probably meant for St William, Archbishop of York from 1140-54' (Saint Albans Cathedral 1952, 14).

Salisbury Cathedral, the Vestry (*Fig. 100*)

The roof is a later restoration, but with early timbers re-used. It is an octagon with straight rafters for eight triangles, and the central pier has an elaborately-carved capital and base, and curved braces. Each of the eight principal-rafters has one curved and stop-chamfered brace, and the wall-posts stand upon stone corbels. There are eight double principal-rafters with trapped purlins and common-rafters.

Wells Cathedral, the Lady Chapel

The Lady Chapel of Wells Cathedral was designed with an irregular polygonal plan comprising an elongated octagon, which was completed in 1319 (Colchester and Harvey 1974, 205-6); the carpenter responsible may have been John Strode (Harvey 1972, 163). The roof is illustrated in Fig. 101, in which the masonry carcase is omitted – but it was finished with a parapet outside the wall-plates shown. The plan figure was tied by two beams from which three, necessarily down-curved ties retained the two semi-hexagonal figures of the ends. The first two tie-beams were mounted on timber-framed brackets, which were fitted with an elegant form of the branching ends noted in the 13th century,

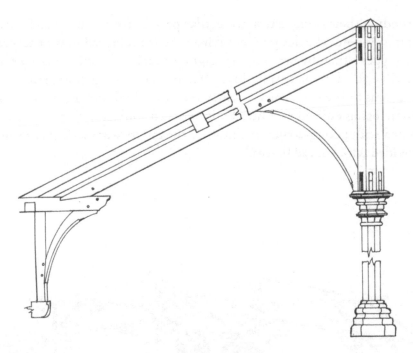

100. Salisbury Cathedral, the roof of the vestry.

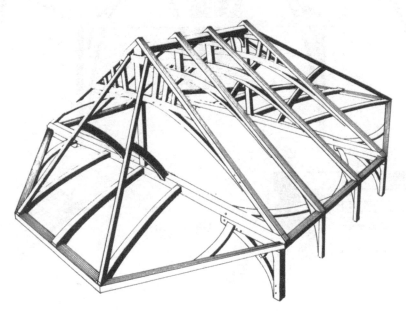

101. Wells Cathedral, the high-roof of the Lady Chapel.

extending their tying action to regular points around the plan-figure. From this point the designer combined the main ingredients of several roof-types – thereby proving the origins of each to be substantially earlier than the date of this building. He used two king-posts from the tie-beams to the ridge-piece, which rests on a high-collar, the normal collars being curved and fitted with curved soulaces. The ends were hipped with raking-struts, and most of the rafters were wall-pieced and provided with curved braces.

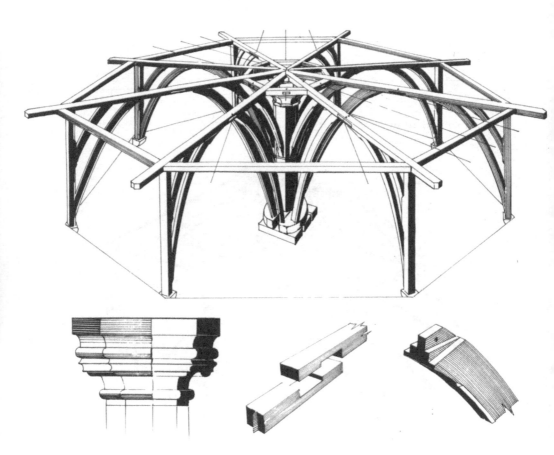

102. Salisbury Cathedral, the roof of the chapter house.

Salisbury Cathedral, the Chapter House

Salisbury's octagonal chapter house was built by Richard Mason *c.*1275 (Harvey 1974, 239). It was of the type employing a central pier, roofed flatly, but with enough slope to drain the leaded surface. The timber roof is shown in Fig. 102, in which the masonry carcase is omitted. The central stone pier was extended by a timber pier of octagonal section mounted on the five heavy pads illustrated. It is a remarkable fact that, at this initial point, the oak pier could have had no vertical stability, indicating an almost simultaneous assembly of the whole complex. The pier was given an elaborately-carved capital, as illustrated. Each of the eight principal-rafters was supported by two curved and stop-chamfered braces from the central pier, and two from the wall-posts that stand upon stone corbels set within the angles of the shell. The concentrically octagonal ring of timbers between the principals con-stitute purlins, and three rafters were chase-tenoned into each sector. The scarf used for the plates is shown below. Chase-tenons with both shoulders spurred were used at all points where the arched braces meet the principal rafters, as shown enlarged.

York Minster, the Chapter House

The chapter house at York Minster is dated to between 1286 and 1296 (Harvey 1974, 246); Dr. E. A. Gee considers its vestibule to have been in existence in *c.*1250, and its octagonal chamber to be of *c.*1280 (verbal information, 1972). The masonry carcase has been altered and height-ened at the time that a stone trabeation was built between the upper parts of its buttresses and the angles of its plan. The timber roof may also be of two 'builds', because its framing in the horizontal plan is dou-bled at the eaves level, but if the lower part of the roof were proved to predate the raising of the eaves, this part must have been raised at the same time. The clear internal span of the building is 58 ft. and the built beams actually span 64 ft., which is the approximate height of the cen-tral mast – the whole work being a relatively obtuse spire. There was no central pier to support vaulting, and the timber 'vaults' are suspended from the spire.

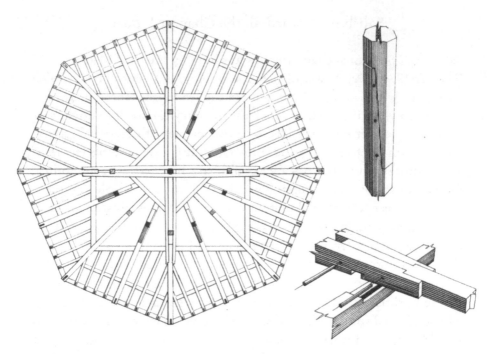

103. York Minster, plan of the chapter house spire at the base.

The two built transoms were formed from pairs of long oaks and for-cibly bent inwards at their ends, where extending single timbers were clasped between them, the whole assembly being trapped by the form of its crossing-joints where passing over the timbers of the square void – to which shorter clamping timbers were subsequently added. The plan of this 'floor'-stage is given as Fig. 103, with a detail of the end clamps. The posts which form a ring within the square void are shown alternately hatched or black, to indicate their two differing functions: those shown black descend to support the vault-rib timbers, and the hatched ones are cant-posts rising within the spire. The pattern, in plan, of the 'vault' is given as Fig. 104. The central spire-mast is scarfed together, possibly to facilitate the assembly. It is probable that the double octagon formed by the wall-plates, with which the eight framed brackets are structurally integral, was first assembled. These octagons of wall-plate must have been closed simultaneously, because all have pairs of single-tenons to secure the horizontal top timbers, which converge on the centre of the

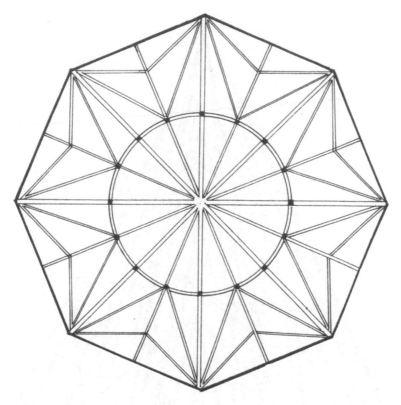

104. York chapter house spire, the pattern of the vault ribs.

plan. From that point, with the lower section of spire-mast in place, the assembly of the floor of radiating joists apparently proceeded, the mast having been steadied temporarily by some simple means. The full complexity of the spire framing is illustrated as Fig. 105, which shows clearly the structural break at eaves level; the lower part of the work was executed with open notched lap-joints, of which there are very few in the spire frame itself. The scarfing of the mast-sections was by means of stop-splayed joints with tongues and grooves, as illustrated in Fig. 103, briefly advocated during the reign of Edward I (Hewett 1980, 265 and 271). It is tempting to theorise at great length upon the structural behaviour of this forest of timbers, and the drawing provides sufficient basis for those who are tempted to the exercise.

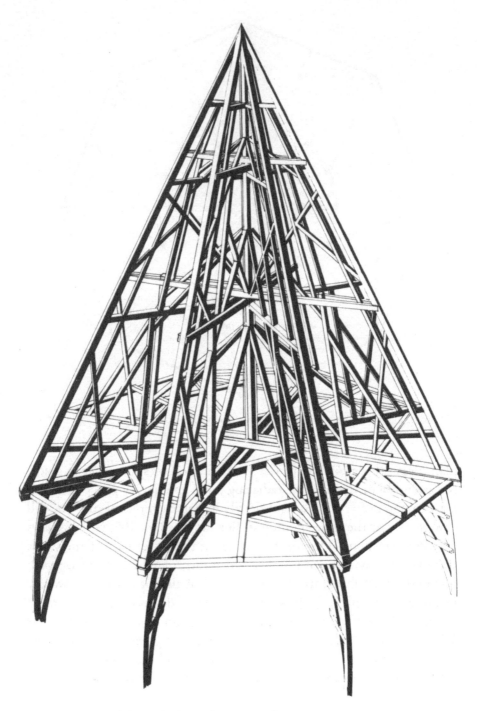

105. Perspective of the York chapter house spire framing.

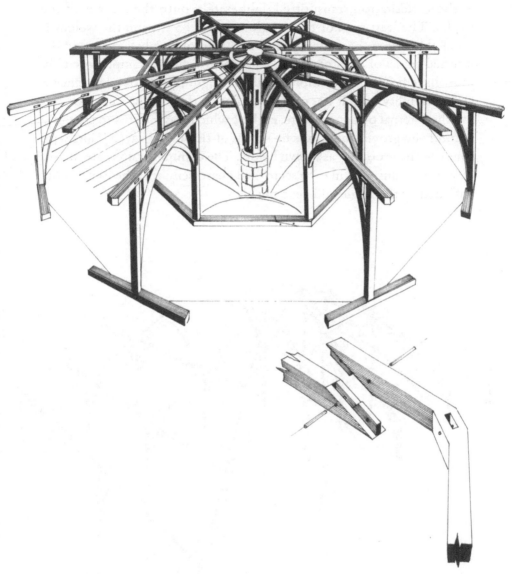

106. Wells Cathedral, the high-roof of the chapter house.

Wells Cathedral, the Chapter House (*Fig. 106 and Plate 10*)

The most spectacular essay in roofing a vaulted polygonal plan is that of the chapter house at Wells, built between *c*.1286 and 1307 (Colchester and Harvey 1974, 205). As in most instances at Wells, the weight was without hesitation transmitted deliberately onto the crowns of the vaults. The least span of the octagon is over 55 ft., and the weight is divided between the central pier, the vault-crown at half-span, and the internal angles of the stone shell. As in the Salisbury example, assembly must have been almost simultaneous throughout the complex, since it lacked stability until it was complete. The scarfing of the timbers forming the internal octagonal plate, each of which was selected to contain a natural and grown angle, was by means of through-splayed joints with tongues and grooves – as shown below. The common-rafters are, in this case, numerous – 12 in each sector, all chase-tenoned, the longer ones stiffened by the single mid-span purlin.

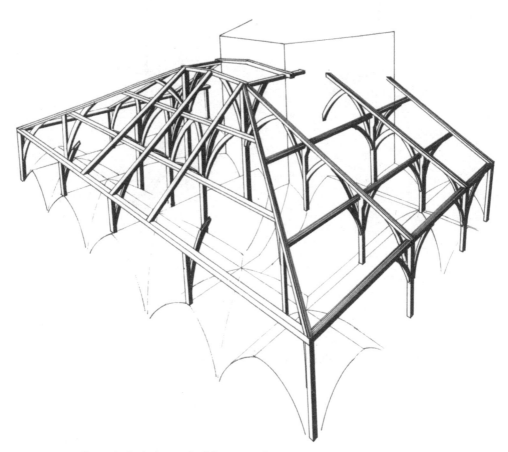

107. Wells Cathedral, the roof of the retro-choir.

Wells Cathedral, the Retrochoir

The Wells retrochoir is approximately square in plan, and appears to have been built outwards from the completed Lady Chapel, begun *c*.1329 and finished by *c*.1345, the work being attributed to William Joy (Harvey 1974, 176). This has a remarkable roof that is in essence fully hipped; it covers a complex of vaults and – as frequently occurred at Wells – the craftsman did not hesitate to gather its weight and transmit it directly onto the crowns of the vaults. As a result of this no less than 14 posts support the system of purlins. This is illustrated by Fig. 107. In this work there are posts from which no less than *six* curved braces spring at the same level. This jointing system may relate the roof to that of the northern transept of Tewkesbury Abbey, and assist with the dating of the latter.

Exeter Cathedral, the Timber Vaults of the Transepts

The timber vault of the south transept is illustrated in Fig. 108. The date of this is, I am assured, absolutely precise: 1317 (Dr. J. H. Harvey, personal communication of 25 January 1973). As shown in the diagram, each rib-timber is supported by two vertically-placed scantlings tenoned at top and bottom; this appears to constitute a development from the diagonally-supported ribs at St Albans, which has an older timber vault. The floor from which this vault is hung must date from the same building operation since it is framed into a square, central void, through which the vault's crown rises. The profiles of the ribs cannot be seen from above or beneath.

Ely Cathedral, the Octagon and Lantern

There can be no doubt that of all the polygonally-planned cathedral buildings peculiar to England, the octagon and lantern of Ely Cathedral must have been the most spectacular. It remains so at the time of writing, but this is in truth illusory, for little of the original *structure* survives. The central tower of the great church, built between

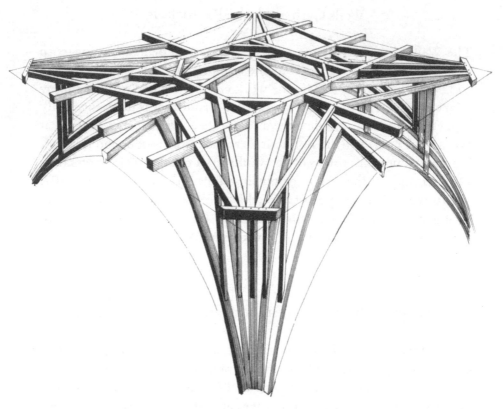

108. Exeter Cathedral, the timber vault of the south transept.

1083 and 1106 collapsed and destroyed the choir on 12 February 1322; under the Sacrist, Alan of Walsingham, the existing crossing lantern and new choir were built – and completed by 1340. The Sacrist's rolls concerning some of the years of this great building operation survive, and have been translated and published, but it has been clearly stated that 'they give very minute lists of the materials purchased and used ... but no hints of any scheme or plan on which the building was proceeding' (Chapman 1907, I, 14). The rolls do prove, however, that William of Hurley – *Magister Carpentarius* to Edward III – 'had a definite and continuous authority in the building' (ibid., I, 45). Several other carpenters are named at various times, and among the earliest was a Master Thomas who was paid 6s 8d, plus 2s travelling expenses, for the 'erection of a great crane for lifting heavy weights' – and it is known that the carpenters inherited, complete and in position, all the scaffolding used by the masons for the octagonal shell. The stonework was finished and the carpentry begun in 1327/8, the relevant roll stating; *Quod quidem opus usque ad superiorem tabulatum per vi annos consummatum anno Domini MCCCXXVIII. Et statim illo anno illa artificiosa structura lignea novi Campanilis* (ibid., I, 35). The roll for 1355 shows that the heaviest expenditure was for carpenters and sawyers working on the lantern, under the direction of William de Houk. The great 'vault' beneath the lantern-aperture was painted between 1334 and 1335, and the second 'vault' beneath the lantern roof was painted between 1339 and 1340. We are also informed that the central boss of the upper roof-'vault' was cut by John of Burwell, in 1337 to 1339, and that it is situated 152 ft. 6 ins. above the floor.

One of the Sacrist's entries has misled people in that it refers to the boarding of eight carpenters 'with the servants of the Lord (Sacrist) through nine weeks for the exaltation, or raising, of great posts in the new choir' – 'pro magnis postibus exaltandis'. This has been thought to refer to the eight huge corner-posts of the lantern, but the entry specifies the choir, and its date, 1334/5, is considered too late for the former (ibid., I, 39). The general dimensions are of interest: the internal span of the lantern is 30 ft., and the circuit of its sides 160 ft., while the internal span of the octagon-plan at ground level is about 75 ft. The total cost of the work was £2, 400 (ibid., I, 68). In July 1757 the architect James Essex surveyed the cathedral and reported to the Dean and Chapter that, among other things: 'The prodigious quantity of Timber and Lead of which it is composed, was at first supported by 16 pieces of Timber only, of which number 7 or 8 are *now rotten* and unfit for supports, so that the whole weight is now unequally supported by those that remain sound' (Rowe 1876, 73). This is also misleading because close examination of the 16 shores shows that all have the same patination, and that none were *replaced*. Their feet, however, have frequently been cut shorter – even to the extent of lowering their tops and subsequently bodging the spurred chase-tenons there, which obviously can never be attributed to Master William of Hurley! James Essex also built into the surrounding space a system of vertical and raking-struts, of questionable merit, which is omitted from the drawings as irrelevant.

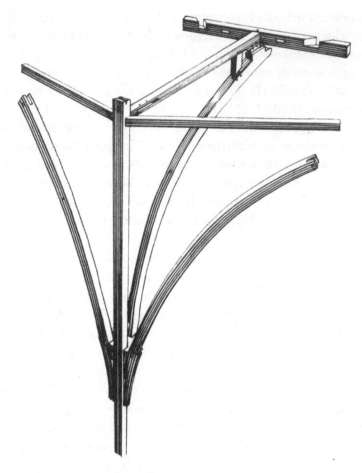

109. Ely Cathedral, the angle post of the lantern.

The method used for construction can be deduced from the fabric, the masonry shell providing certain evidence for the former quantities of heavy timbers that are now absent. As shown in the following series of diagrams, there was no baffling sequence of assembly, such as it is tempting to envisage. The full-height grooves built into the internal angles of the stone shell were fitted with eight great posts, probably measuring 9 ins. by 11 ins., and 48 ft. in length (Fig. 109). These were retained, apparently, by the timber tie-backs which were let into the stone hooks at either side of each channel, and thereafter the side-timbers framing the octagon-plan were inserted between the posts, rendering them immovable. This cannot be proved owing to the ravages of decay, boring insects and restorers, but it is highly probable. Timber imposts for the springing of the numerous vault-ribs were fitted to these posts, and rested, of course, on the upper stone surface of the piers. These remain *in situ*, but have been carved away on their interior surfaces to eradicate beetle, and no longer form a sound standing for the ribs – some having buckled sideways, causing an angularity in the originally straight lines of the 'vault'. The side arches were no problem and were probably fitted without delay, these having a lodging on a projecting stone course throughout their arcature. The bracket-like assemblies

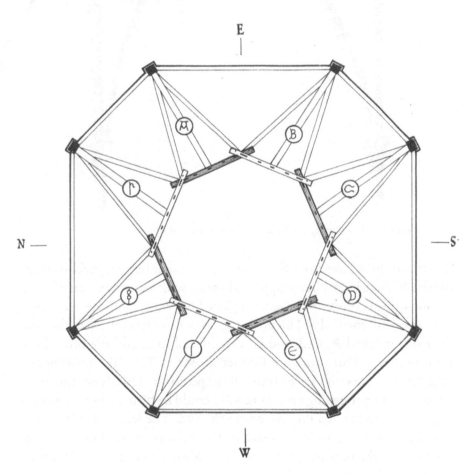

110. Ely Cathedral, plan of the floor of the octagon, showing the letters scribed in the timbers.

111. Ely Cathedral, assembly of the inner sill of the octagon.

illustrated must have been fitted thereafter; the inherited great scaffold (with additions) must have supported these, which examination of their crossing-joints proves to have been placed in two sets of four – perhaps almost simultaneously. The plan, Fig. 110, illustrates the fact that the four sills marked A, C, E and G were placed first, and the other four let into them. This process is further shown in Fig. 111, showing the placing of the second set of four sills in position. This done, the joists steadying the projecting ends of the sills could be fitted, when a perilous degree of stability would have been achieved. The letters used to identify the timbers cut to fit into each of the eight sectors are illustrated in the form of the originals, perhaps cut by Master Hurley. I am indebted to Dr. O. Rackham for the information that all were cut with two scribes, one for straight cuts and the other for inch and a quarter circles

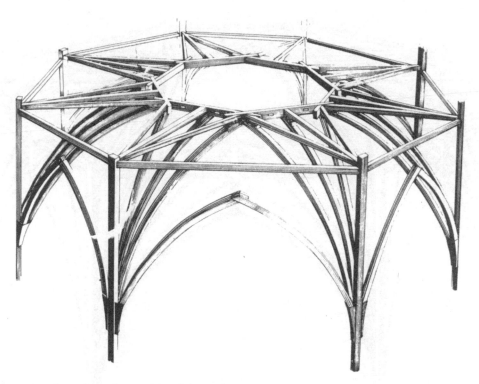

112. Ely Cathedral, assembly of the vault.

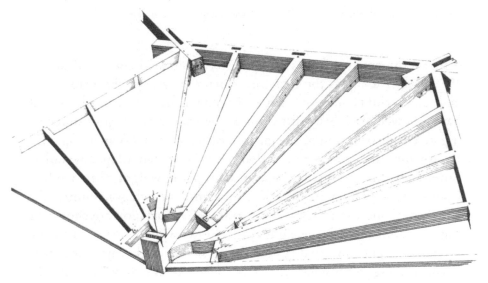

113. Ely Cathedral, sector of the 'floor' of the octagon.

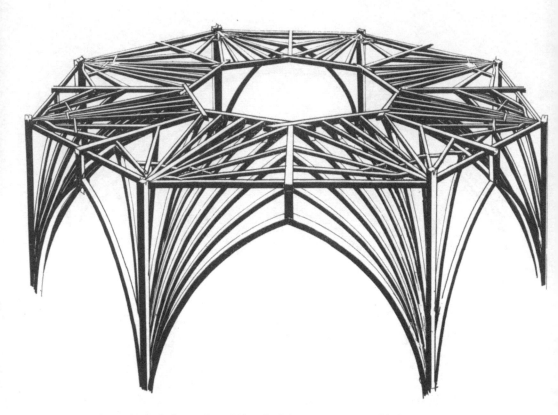

114. Ely Cathedral, the vault and 'floor' of the octagon, assembled.

– hence the curious forms of the letters (personal communication, 1979). Fig. 112 shows the structure when the first 24 ribs were fitted, and some stability achieved, whereafter the many tierceron-ribs would have followed, each being fitted simultaneously with the joist supporting it. The distribution of the joists in each sector is shown in Fig. 113, from which it must be noted that every joist had a lap-dovetailed outer end-joint, to resist withdrawal after the weight of its rib had been suspended from its inner end – against the octagonal inner sill. The 'vault' and the integral floor to the 70 ft. octagon are shown in Fig. 114, above which the eight corner-posts would have risen a considerable height; into that system the 16 shores could easily be fitted, having passing-joints where they cross major joists – shown in Fig. 115,

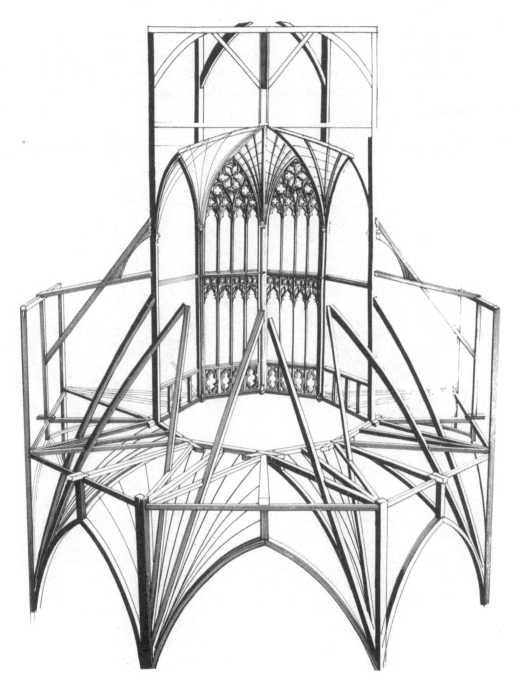

115. Ely Cathedral, the framing of the lantern.

suitably simplified. The same diagram illustrates the manner in which the lantern's angle-posts could be reared against the shores, resting on the sills to which they were tenoned. (This has been simplified by omitting a large number of timbers, and the pinnacles which surmount the lantern.)

These enormous angle-posts are not all, as has been thought, in a single piece. Fig. 116 shows that some of them were split lengthwise,

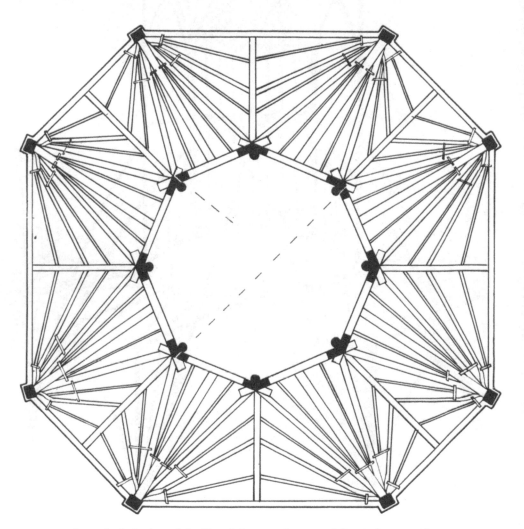

116. Ely Cathedral, plan of the 'floor' showing division of the angle posts of the lantern.

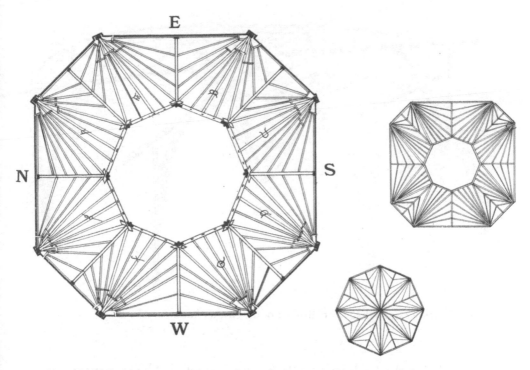

117. Ely Cathedral, the patterns of vault ribs in the upper and lower systems.

in order that the lantern might be erected in position first as a half-octagon, then completed by adding the other two quarter-octagons.

The patterns of vault-ribs in the upper and lower systems are shown in Fig. 117, which illustrates the irregularity by which the unequal outer octagon was converted into the inner and regular octagon of the lantern. The smaller pattern at bottom right is that of the upper 'vault' forming the ceiling beneath the floor of the bell-chamber in the lantern's top space.

Worcester Cathedral, the Chapter House (*Fig. 118*)

The Worcester chapter house has a circular plan, and is dated so far as the carcase is concerned to *c.*1120 (Harvey 1974, 212, 244). The timber roof is believed to date from between 1386 and 1392 – the work being

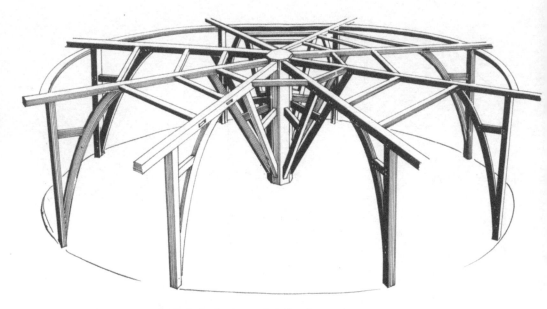

118. Worcester Cathedral, the high-roof of the chapter house.

recorded in the Cellarer's Rolls for the Cathedral. The carpenter was named Hugh, and he received a fee of 52s a year plus 26s for food. I am indebted to Dr. J. H. Harvey for this information. The whole possessing a central pier the design, again, extended the pier with timber, and employed eight principals radially disposed with three straight purlins between each pair. Each principal was stiffened by tied straight braces at its centre and by strutted curved braces at its wall-posts.

Lichfield Cathedral, the Lady Chapel (*Fig. 119*)

This shows the roof which surmounts the Lady Chapel, built *c.*1320-6 by William of Eyton, but which must apparently date from the massive restoration of 1661-9 by Sir William Wilson (Harvey 1974, 229). This is built as a three-sided half-conoid roof leaning against a king-post frame, fitted with queen-posts and a straining-collar. The hip-type rafters of the apse are forelock-bolted to the head of the king-post. Like the other Lichfield roofs resulting from this general restoration,

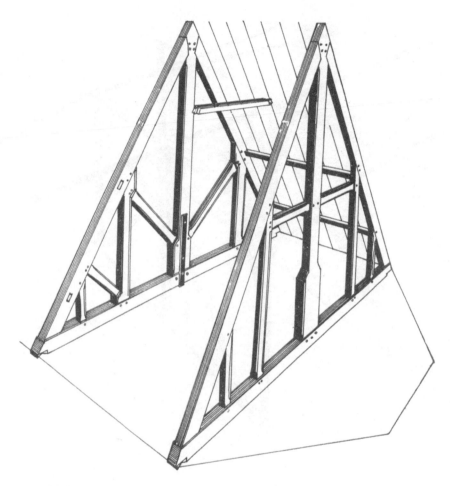

119. Lichfield Cathedral, the roof of the Lady Chapel.

it is massively framed in heavy, chamfered timber, and indicates a very high order of costs. The common-collars in the bay next to the apse end, fitted in addition to the two side-purlins, show a closer affinity with medieval roofing than might be expected at this late date. Critically, the timbers are overweight for their purpose, like those at St Paul's.

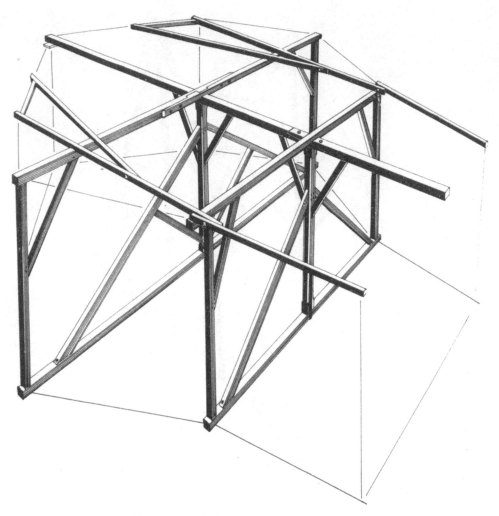

120. Westminster Abbey, the roof of the south-east transept.

Westminster Abbey, the South-East Transept (*Fig. 120*)

The roof consists of portal frames tied at their bases, that were stiffened by centre posts braced both down to the portal ties and up to a central purlin. The roof is hexagonal with lateral purlins jointed to each other with forelock-bolts. The date is about 1650-1700.

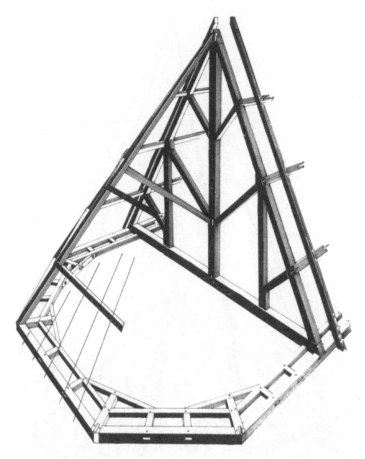

121. Pershore Abbey, Worcestershire, the roof of the apse.

Pershore Abbey, Worcestershire, the Apse (*Fig. 121*)

Built in three sides, this example leans against a king-post trussed frame like the previous one, but is far more elaborate, and its slender timber is disposed in a more sensitive manner, and with more advantage. The date of this structure, which is evidently a replacement, is uncertain but likely to be during the 18th century, if compared with others known to be of that period, such as the Hawksmoor roofs at Beverley Minster.

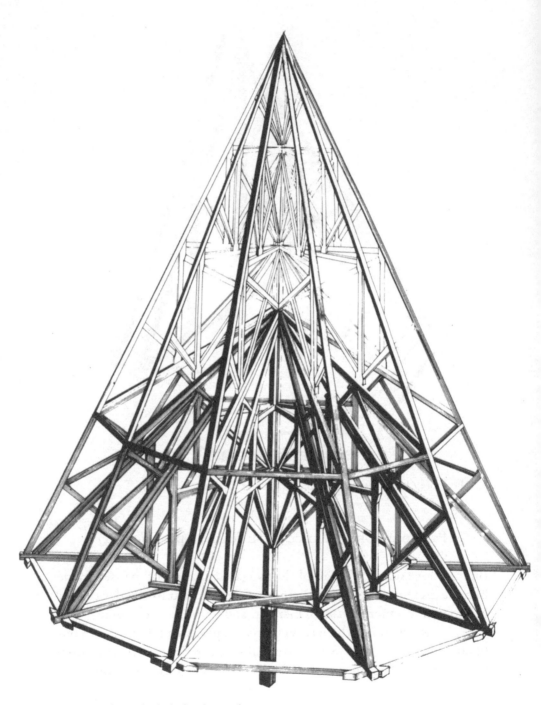

122. Lincoln Cathedral, the chapter house spire.

Lincoln Cathedral, the Chapter House

The first roof is the 'gambrel'-type decagonal structure that is heavily shaded in Fig. 122, which was built by James Essex in 1761-2. The second build, which restored the roof to a full pyramidal form, shown unshaded, was executed in 1800 – architect unknown (Venables 1883, 397). The lower part is of puzzling complexity and, although entirely of softwood and mainly assembled by ironwork and forelock-bolts, is the product of a master. Had better timber been available, it would have been assured of considerable durability. The pine used, however, is not capable of enduring the stresses to which it is subjected, particularly where it forms the radiating tie-beams that are much cut about to form their peripheral joints with the wall-plates. Why James Essex should have persisted in the use of forelock-bolts, long after threaded bolts with nuts had become available, is difficult to determine; but an aim for low costs could be the answer – since in terms of efficiency there may be little difference.

The basic unit of the design, visible when viewed on any one of its diameters, is of two superimposed queen-post assemblies set inside a pitched roof having a king-post. Of this, the principal rafters are doubled and a king-post truss with raking-struts is mounted on the outer rafter – producing the two pitches of the 'gambrel' roof. While this is

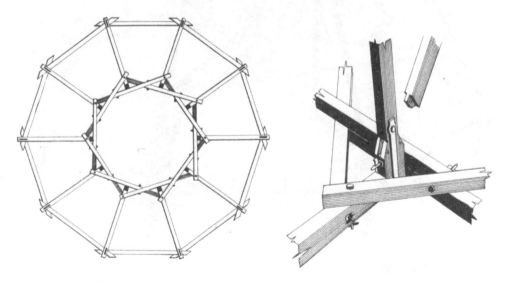

123. Lincoln Cathedral, assembly of the ring-beam of the chapter house spire.

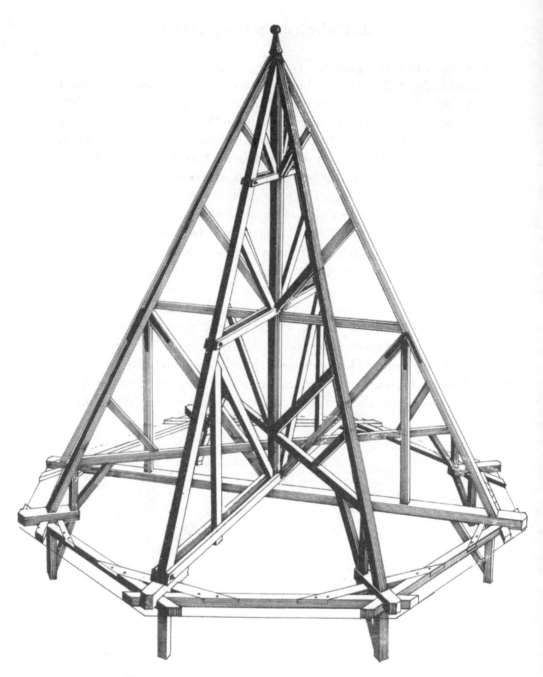

124. Southwell Minster, the roof of the chapter house.

a combination of two known principles the construction of the essential 'ring-beam' that secures the inner ends of the 10 radiating ties may
be the architect's invention, and is so ingenious as to merit illustration
(Fig. 123). The plan attempts to explain the method of its assembly, and
the perspective detail shows the assembly of the ties and the 10-pointed
star formed by the beams. Words will no further clarify these drawings,
which will be found intelligible if protractedly studied. The decagonal tie-beam so formed has proved until this time to be adequately
strong, although of pinewood, and is subjected to both radial extension
and shearing stress. The wall-plates have also proved adequate despite
the fact that earlier examples in lesser structures (the spires of parish
churches) have been made in oak; but the tie-beams, where they cross
the crossings of those plates, are likely to fail – mainly owing to the
inadequacy of pine for such purposes.

Southwell Minster, the Chapter House (*Fig. 124*)

This is by Ewan Christian, who restored the Minster in 1868, the work
continuing until at least 1886.

CROSSINGS AND RETURNS

Salisbury Cathedral, the Return of the North Porch and North Aisle

The conjunction of the ridged roof of the north porch with the lean-to roof of the north aisle, at Salisbury, is the original – and is preserved intact. This may be dated to the middle years of the nave's building period, 1237 to 1258, or *c.*1248 (Harvey 1974, 239). That such a conjunction was intended from the start is proved by the valley-rafters that were framed into the northern lean-to roof at this point; the couples of the parvise roof were stepped with tenons, up the valley-rafters – as illustrated by Fig. 125. This drawing includes some peculiarities also, such as the arcaded inner wall-plate of the triforium roof.

Salisbury Cathedral, the Return of the North Choir Triforium and the Eastern Triforium of the Great Transept

The roofing of the angle formed by the junction of the northern choir triforium and the eastern great north transept of Salisbury is original to that part of the fabric, which was built between 1225 and 1237 so far as the choir is concerned, and between 1237 and 1258 for the great transept (Harvey 1974, 239). It cannot be estimated in what year of this operation

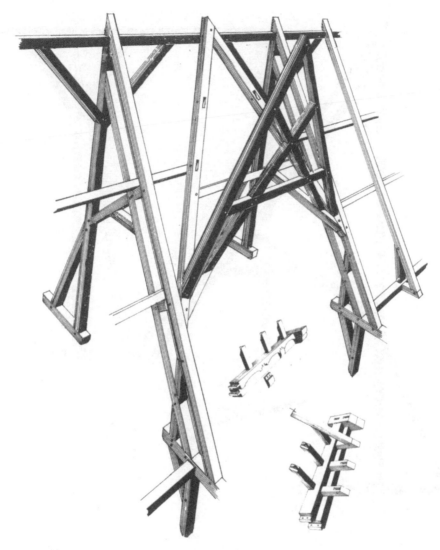

125. Salisbury Cathedral, the return of the north porch and north aisle.

the angle was completed and roofed – but clearly the years spent upon the vaulting of the aisles beneath the high-roofs must be deducted from the final date. This work is shown in Fig. 126, with the common-rafters indicated by single lines in order that the main framing remains visible. The logical bisection of the angle was the first step, the heavy transom occupying that position, and the stabilising of the end inside the

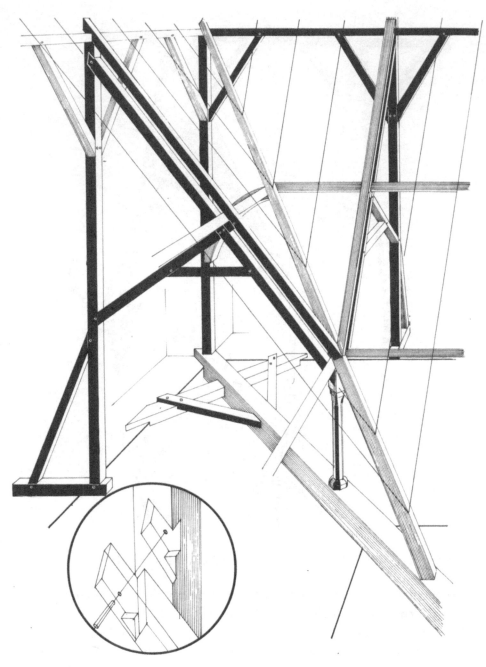

126. Salisbury Cathedral, the return of the north choir triforium and the eastern triforium of the great transept.

return was assured by the system shown. A series of posts approximating to a timber arcade was then set up, standing upon soles and united by the top, or head-plates, when the valley-rafter was fitted and mounted on top of the carved octagonal post – from which under-rafters sprung and were jointed with secret notched laps into the flanks of the first posts. The raking-struts had their compressive moment deflected from the rears of the soles, and were halved through the under-rafters in order to support the upper of the two purlins – when the whole unit was fitted with a collar in compression. It is probable in view of this complexity that the transverse units were set up, complete in all but their purlins and rafters proper, and then secured by fitting the top-plates onto both post-heads and braces. After this the valley-rafter could be laid *over* the purlins, and the jack rafters fitted. An 'X-ray' view of the secret notched laps is shown below, and other details in Figs. 228 and 273.

Lincoln Cathdral, the Choir Crossing

This has evidently been rebuilt from inside to its valley-rafters. It is possible that the original may be hypothetically restored since several jack-rafters exist along the lines of all four valleys. What survives today is shown as Fig. 127, as though viewed from the north-west. A detailed study of all notch-lapped jack-rafters at this point might render a reconstruction possible.

York Minster, the Chapter House Vestibule

The roof of the chapter house vestibule at York Minster dates to *c*.1310 (Dr. J. H. Harvey, personal communication, 1973). It was scissor-braced and chase-tenoned; the framing of the return is illustrated in Fig. 128. The angle was bisected by a diagonal tie-beam on which a king-post was set, and down-braced by two under-rafters from which raking-struts propped the principal-rafter of the return. All the 'jack'-rafters of both the 'valley' and the hip were tapered to fit the edges of the principals, and spiked. The wall-plate was a single timber laid centrally along the masons' top-course, and the sole-pieces were housed over it.

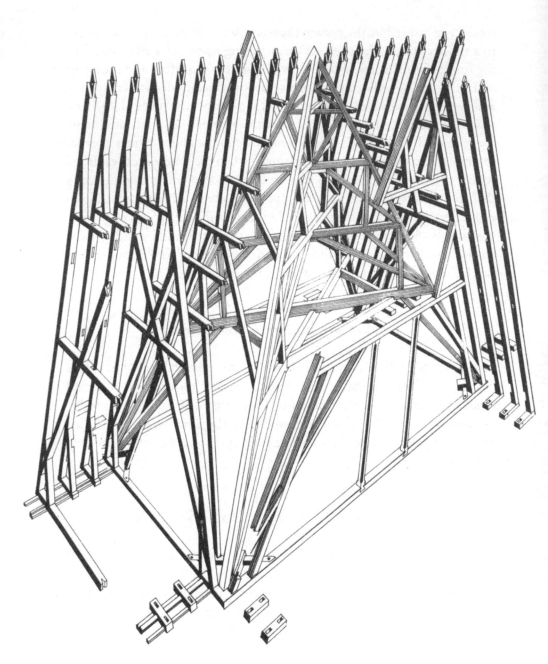

127. Lincoln Cathedral, the roof of the choir crossing.

128. York Minster, the roof of
the chapter house vestibule.

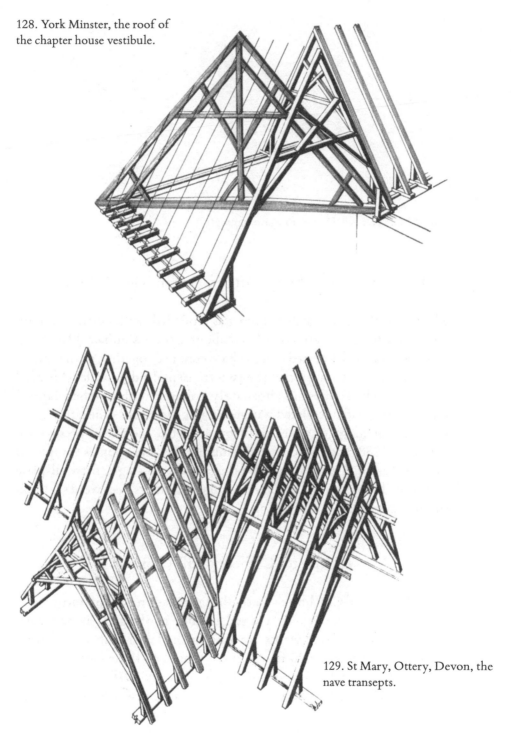

129. St Mary, Ottery, Devon, the
nave transepts.

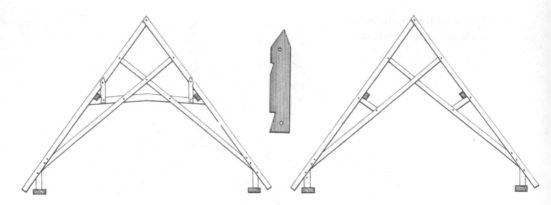

130. St Mary, Ottery, Devon, the nave transepts.

The Church of St Mary, Ottery, Devon (*Figs. 129 and 130*)

The fabric of the collegiate church was modelled 'with much care upon Exeter Cathedral. It was completed about 1342' (Whitham 1956, 8-9). The roof is a scissor-braced design chase-tenoned together, with its main run aligned east to west, and the two transeptal ridges valley-boarded against it. The frames are alternately collared with notched-lapped purlin struts, and uncollared with raking purlin struts – the latter being the essential design of the eastern couples above the Exeter presbytery, begun by 1288, and the former resembling the later couples completed by *c*.1310 as far as one bay west of the crossing. A single, flat wall-plate was provided for the ashlar-pieces, while the rafters' feet were set upon suitably angled courses of ashlar.

Chichester Cathedral, the Cloisters

The cloisters of Chichester Cathedral possess one of the most spectacular returns known in ridged timber roofing. It is not very large, because cloister walks are among the lesser features of great churches, but it spans 20 ft. All three ranges of these cloisters were built between *c*.1400 and 1500, possibly by William Wynford (Harvey 1974, 223). The west range returns eastwards at an obtuse angle; both ranges have 'compassed' open timber roofs, framed with curved soulaces and ashlar-pieces. The pointed arch thus formed turns the corner by means of a 'mitre'. This was effected by bisecting the angle with a principal-rafter couple that was itself framed into an arch, with curved soulaces and ashlars, its arch necessarily flatter than those of the straight ranges, the couples of which all tenon into the bisecting frame at approximately 45-degree angles. The drawing, Fig. 131, illustrates the principle. This elaborate work appears to be chase-tenoned at all points; if it was made in a framing yard and transported to the site for assembly, as was frequently the case, this was an astonishing achievement.

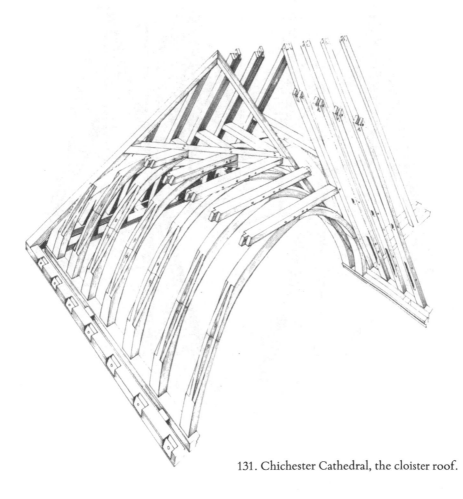

131. Chichester Cathedral, the cloister roof.

Salisbury Cathedral, the Choir Crossing

The crossing roofs of the Salisbury choir transepts and nave must have been framed in some manner that equalled the directness, and surprising thoroughness, of the few pieces of original roofing that survive there. James Wyatt was responsible for many alterations in the period 1787 to 1793 – together with some demolition works (Dale 1956, 102-111) – and it is to his time that the existing roofs belong, one puzzling timber being inscribed 'Jas Larkim 1780'. His framing for this intersection is illustrated in Fig. 132. It is impressive to view mainly by virtue of its

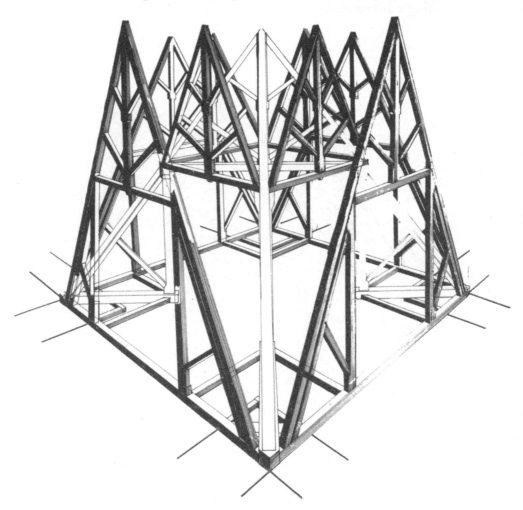

132. Salisbury Cathedral, the roof of the choir crossing.

size, having a choir span exceeding 40 ft., and an approximately equal height. This work is too complex, when seen from an angle, to show all the details, so some are lightly indicated for the sake of clarity. The tie-beams of the choir and transepts were laid in a square and framed into dragon-beams that reach inward. On each of these tie-beams was built a queen-post frame with raking-struts to its underrafters; king-posts were set on the collars, and the eight principal-rafters added. On the internal ends of the dragon-beams four further posts were mounted and collared, being provided with diagonal straining-beams to prevent their sagging towards the centre. Each side of this interior square mounted a king-post truss matching the first four; and from the corners of the first and basal square rose the four valley-rafters which rest on the corner-posts of the internal square. They are strutted by a central king-post mounted on the crossing, diagonal strainer-beams of the internal square – which is at collar height. The pitched planes of these roofs were formed by a series of common purlins.

TOWERS AND SPIRES

Canterbury Cathedral, the South-East Spirelet

The small timber-framed spire surmounting the south-east stair turret of Canterbury Cathedral, illustrated in Fig. 133, is of considerable interest. The historical background has been fully published (Hewett & Tatton-Brown, 1976) but must here be summarised. The choir and south-east transept were gutted by fire in 1174, and the evidence that the flames destroyed the capping of the south-east tower is clearly discernible, so the existing spire must date after that event. The account written by Gervase the monk of Canterbury implies that the rebuilding of the gutted eastern parts was completed by 1184 and no *major* building works were in hand from then until 1220, apart from the glazing (Willis 1845, 32-62). The spire framing relies upon notched lap joints of the secret category, and is considered to date from 1184 or later. The tower is of square plan and contains only the spiral stairs which ascend the full existing height, leaving the original termination doubtful, but now there is only the spirelet. This is built inside a parapet, is octagonal, and is framed upon two transoms which meet at the centre, of which only the east-to-west beam is continuous. Four short timbers tenon into these close to their intersection, and from their mid-points radiate four short ties – bringing the total to eight, one to each arris of the spire. This curious setting-out presents one facet of the octagon to each face

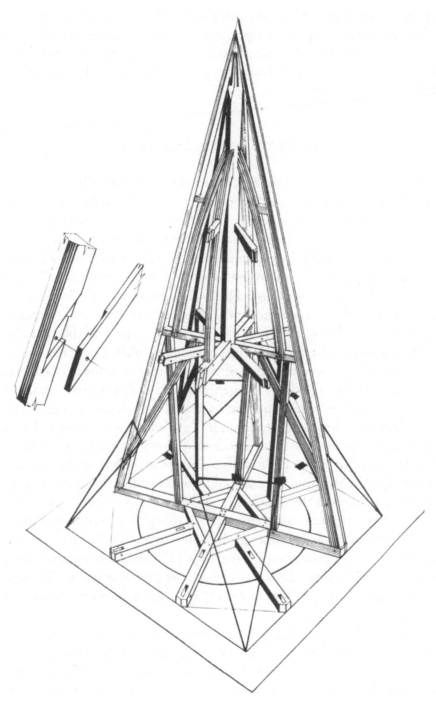

133. Canterbury Cathedral, the south-east spirelet.

of the square tower. A star of eight radial collars is framed at one-third of the height, and supported upon eight posts from the first beams. The arms have soulaces as illustrated, with secret notched laps at their conjunction with the rafters. The spire-mast is stepped on the collars and braced down to them.

Salisbury Cathedral, the Spire

The central tower of Salisbury Cathedral mounts the tallest surviving spire in England, with a published height of 404 ft. excluding the iron cross, and an internal floor area of 32 ft. square. The carpentry in this work is everywhere of the surprisingly direct and remarkable type that is noticeable in the high-roofs and the triforia lean-to roofs. The first timber-framed floor is mounted over the crossing-vault, at a height of 83 ft. above the nave floor. This is illustrated in Fig. 134. It is not boarded now, but was evidently designed to be so covered. It depends upon the availability of very long oak trunks – and was substantially stiffened by the bracing beneath the bridging-joists. Forty feet above this, and a little above the ridge of the nave-roof is mounted a gallery, illustrated in Fig. 135. This is ingeniously framed; the great angles of the square void were first spanned by beams at 45 degrees, onto which were laid joists that bisected the angles, into which were chase-tenoned three common-joists to form a walk-way along each side. A forelock-bolt was used to secure the first two timbers at their crossing; and the internal common-joists were braced from beneath – the braces springing from pockets in the masonry, and crossing each other during their ascent. The hand-rails built around the gallery thus produced seem to be the earliest 'trussed' timber girder construction yet known, and it is of interest that all the compressive diagonal struts meet square, hewn, outsets on the posts; these later became square abutments. Though much repaired this gallery is mainly of original timbers, and of its initial design.

A little below the gallery is a system of iron ties that appears to be an original feature. Fig. 136 illustrates one-quarter of its plan area. The whole is wrought, and secured by forelock-bolts. Concerning this it was stated in 1774 that: 'It seems as though the architect himself was

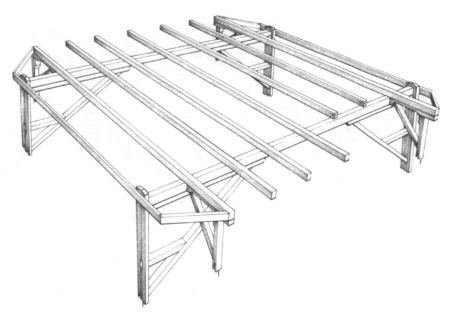

134. Salisbury Cathedral spire, the floor over the crossing vault.

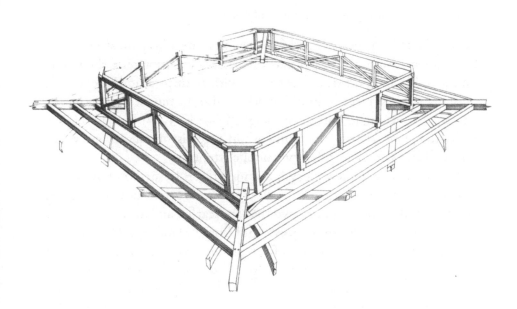

135. Salisbury Cathedral spire, the gallery.

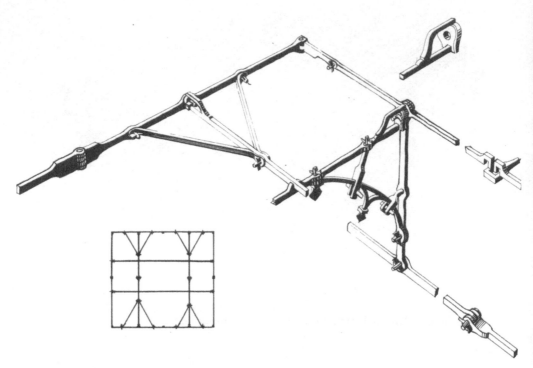

136. Salisbury Cathedral spire, the system of iron ties.

not without his jealousies and fears. At first, he adds a most excellent bandage of iron to the upper part of the arcade, embracing the whole on the inside and outside of the tower, with an uncommon care: this is, perhaps, the best piece of smith's work, as also the most excellent mechanism, of anything in Europe of its age' (Price 1774, 14). It is, perhaps, a little strange to describe this ironwork in a treatise upon carpentry, but it is too little known and deserves public attention.

Forty feet above the gallery is a simple floor, framed on two bridging-joists of the necessary 32 ft. clear span, plus their bearing lengths at either end, which support two that are scarfed and halved into their upper faces, with common-joists and board cladding. This is illustrated by Fig. 137. Upon this a four-posted structure 33 ft. high was built, designed to distribute its weight and that of the superstructure evenly betwixt the masonry carcase and the bridging-joists. The posts were secured by stirrup-irons and forelock-bolts. Three feet above this there is a framed timber floor, illustrated in Fig. 138. At this height

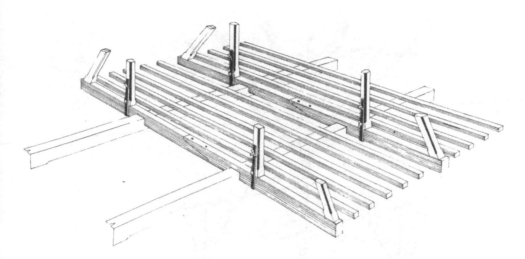

137. Salisbury Cathedral spire the floor above the gallery.

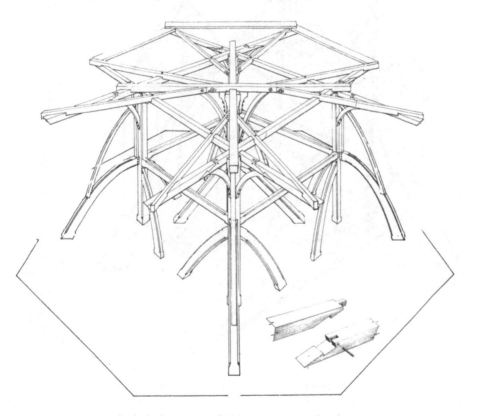

138. Salisbury Cathedral, the spire scaffold above parapet level.

139. Salisbury Cathedral spire (left) the next three stages (right) the remains of the ironwork from which the scaffold was suspended.

the squint-arches that reduce the square to an octagon affect the plan; the floor is framed to form an octagon, but not to support the spire-scaffold, whose base timbers are mounted on the upper transoms of the square tower-like internal structure.

Fig. 139 shows the three subsequent stages, in order of ascent, the lower being the second above parapet level, and one of the most remarkable works of carpentry known. Finally, at the side is shown the remains of the original iron 'machine', by means of which the weight of the whole scaffold assembly (down to the point where the central mast fits over the transoms with four toes) could be suspended from the iron cross, which penetrates the capstone.

At the unknown date after 1320 when this work was begun the master concerned introduced the scarf joint illustrated in Fig. 138. This seems to have been the first scarf ever to possess a bridled abutment; the reasons for its use at this point, or in this construction, are not yet determined, but its effect upon the subsequent development of scarfing was remarkable. The spire was probably finished before 1387 (Harvey 1974, 215-6).

Lincoln Cathedral, the Roof of the Crossing Tower

The latest of the medieval roofs at Lincoln is that which tops the crossing tower; this was completed in its present form by Richard of Stow between *c*.1307 and 1311 (Harvey 1974, 231). The framing is illustrated by Fig. 140 showing its octagonal plan, doubtless for the base of the leaded spire formerly built there, attaining a reported height of 525 ft. until blown down in a storm of 1549. It was based upon eight great joists extending from north to south, fully 45 ft. in length, which are met by nine shorter ones; the angles are bisected and filled with trimmed joists. The whole assembly is mounted low within the tower where an octagonal plate is mounted on stone corbels, from which numerous posts with jowls support the upper plates, and are stiffened by doubled braces and descending ties such as characterise carpentry of the reign of Edward I. Fig. 141 illustrates one angle of this framing from below, showing the stone squint-arches by which the plan was reduced to an octagon – the whole resembling the framing and masonry at Salisbury.

A floor at Lincoln Cathedral that is similarly framed is shown by Sir Banister Fletcher in the first edition of 1896. It has not as yet been examined, but is probably of similar construction.

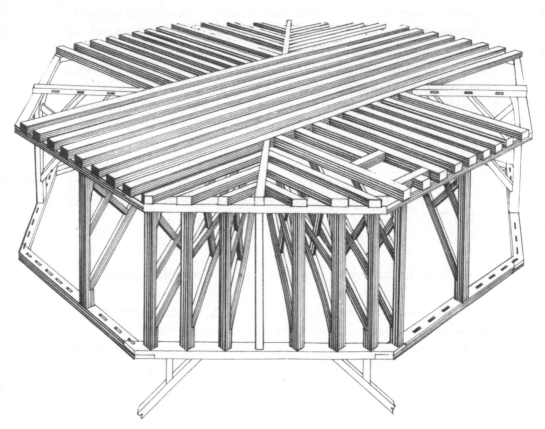

140. Lincoln Cathedral, the base of the former spire.

141. Lincoln Cathedral tower, the stone squint arches and timber structure
from below.

Winchester Cathedral, the Roof of the Crossing Tower

The crossing tower of Winchester Cathedral was built between *c.* 1108 and 1120, this period including that for the construction of the north and south bays adjoining (Harvey 1974, 243); this was almost certainly surmounted by a lead-clad timber spire built in 1200. *Anno MCC inchoata est et perfecta turris Wintoniensis ecclesiae* (Annals printed in H. Wharton, *Anglia Sacra*, 1691, 304). I am indebted to Dr. J. H. Harvey for this information, concerning which he observed: 'No stone tower could have been begun and finished in a single year, but *turris* is a word used of timber spires in medieval sources; this record therefore gives a precise date for a lost timbered work, from which some framing might survive'. The existing timber roof to this tower is illustrated by the diagram, Fig. 142, which clearly shows that no framing relating to the former spire does survive – unfortunately. The four tower piers of the crossing form a 50-ft. square, and the floors and roof are equally large.

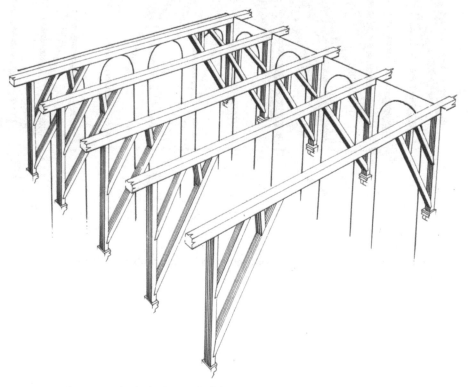

142. Winchester Cathedral, the roof of the crossing tower.

Pershore Abbey, Worcestershire, the Roof of the Tower

The Abbey church of Pershore has its Norman tower surmounted by a 14th-century lantern, the earlier bell tower having been destroyed by fire in 1288. It is considered probable that masons and carpenters were obtained from Salisbury to effect the rebuilding at Pershore (Moore 1961, 7). In 1335 the tower was still ruinous; it is likely that the existing lantern with its square conoid roof was completed in the following decade (P. Barrett, personal communication, 1971). This is framed as illustrated in Fig. 143, with hipped ends within the parapet. It has three king-posts mounting a ridge-piece, and supported laterally by cranked

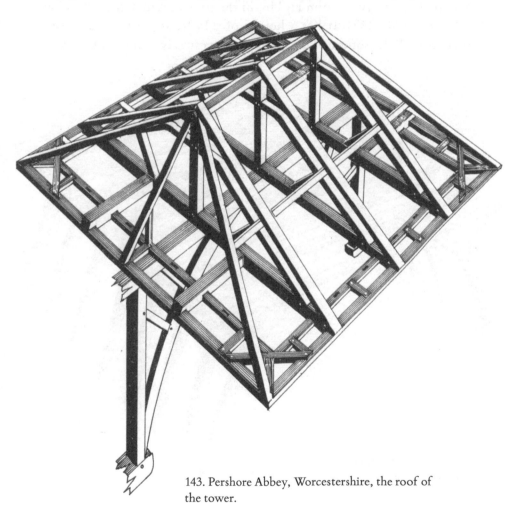

143. Pershore Abbey, Worcestershire, the roof of the tower.

under-rafters with raking-struts. The wall-plating is double, and keyed together, with angle-ties across the corners and dragon timbers for the feet of the hip-rafters. The three tie-beams are stiffened by a massive portal frame beneath, which has curved and doubly strutted bracing, with wall-posts mounted upon stone corbels. All the common-rafters have ashlar-pieces.

Gloucester Cathedral, the Central Tower

The central tower of Gloucester Cathedral was built between 1450 and 1460, possibly by John Hobbs, in the style in which it exists today (Harvey 1974, 227). It has a height of 225 ft. and a floor area of a little over 20 ft. square. Two framed floors survive, one illustrated in

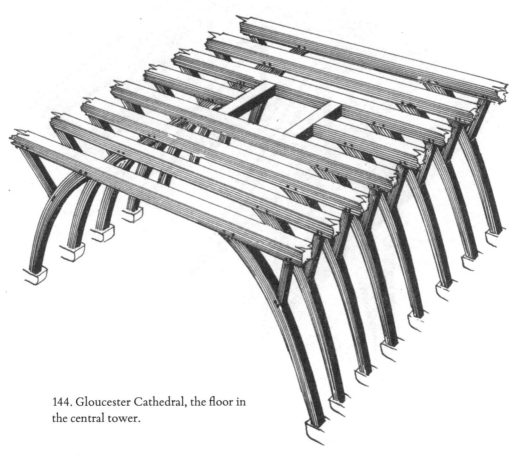

144. Gloucester Cathedral, the floor in the central tower.

Fig. 144. These have eight massive joists supported upon curved struts that spring from stone corbels, each strut being strutted within the void of its spandrel.

Canterbury Cathedral, Bell Harry Tower (*Fig. 145*)

Important new evidence found in 1975 shows that the four tie-beams – each having lap-dovetailed ends of late 15th or early 16th century type – that span the tower from east to west, have quite steeply humped top surfaces which make them unsuitable to lay a floor on, but ideally suited for the slightly cambered roofing of the tower. Apart from this 'buried' roof, there is other internal evidence to indicate that the tower was built in two stages. It is well known that Bell Harry Tower is in fact a brick tower, only faced with stone, and that between 1494 and 1497

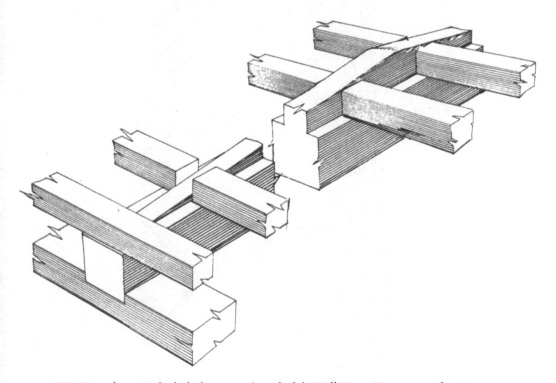

145. Canterbury Cathedral, the original roof of the Bell Harry Tower, now forming an internal floor.

nearly half a million bricks were used in its construction. These bricks are very distinctive and can be clearly seen in the walls of the upper chamber and in the stair turrets. However, as one descends the stair turrets, a string course of stone is reached, and below this much smaller bricks can be seen. In 1486 on Bourgchier's death, John Morton became Archbishop, and in conjunction with Sellinge planned the rebuilding of the central tower as a simple lantern tower, which must have been finished and roofed before 1493 (Hewett and Tatton-Brown, 1976, 133).

146. Norwich Cathedral, the inner frame of the spire.

Norwich Cathedral, the Spire

The spire of Norwich Cathedral reaches a published height of 320 ft. It was designed and built by Robert Everard between *c.*1464 and 1472 (Harvey 1974, 234). J. Adey Repton, writing in 1794, observed that 'there does not appear to have been any wood framing in it. In the late, and the present Dean's time much has been done to keep it in repair, and by putting in framed timbers to preserve it'. The Dean was the Very Reverend J. Turner, from 1790 to 1828 (Pierce 1965, 20). The timber-work comprises 12 frames like those illustrated in Fig. 146, with some variations at apex and bell turret. Mainly of pine, assembled with lap-dovetails and spiked together, the helm-type windlass (Fig. 194) appears to be of the date of this internal framing. It stands, however, upon an

147. Southwell Minster, the base of a spire.

oak-joisted floor that is cambered, and for that reason seems to have been the roof before the spire was added.

Southwell Minster, the Spires

Of the restoration by Ewan Christian, the two spires crowning the twin west towers are interesting; the base of one of these is shown in Fig. 147. Framing in pine, and heavily braced.

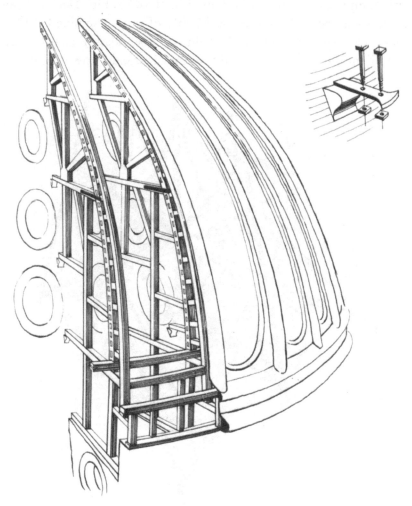

148. St Paul's Cathedral, the framing of the dome.

St Paul's Cathedral, the Dome (*Fig. 148*)

The dome was constructed by Sir Christopher Wren from 1707 to 1708 (Lang 1956, 237-40). There are two domes between which is a brick cone to support the lantern, ball and cross (drawing in Banister Fletcher, 275). The outer dome, covered in lead, is 130 ft. in diameter and 60 ft. high. Six iron chains clasp the structure. The top of the dome is 274 ft. from the ground. The brick cone is 18 ins. thick, and radiating from near its base are 34 oak plates. Framed above these are four other stages of ties braced to the centre cone and the outer timber dome. The curved ribs of the outer dome are all made from two members scarfed together, the scarf being strapped to a radiating tie. It is most probable that the ribs were steam-bent to shape – a technique known to have been used in ship-building at that time. The spaces between the ribs were linked by multiple horizontal purlins, no common rafters being employed. This technique had already been used in the nave roof of Tewkesbury Abbey.

DOORS

Westminster Abbey, the Door called 'Pyx'

The oldest door known at this time, within the named category, is that suspected to have been originally the door to the Pyx chapel of Westminster Abbey, now relegated to a position behind the bookstall in the eastern range of the cloister. This is unlike other known doors, and differs so greatly from such surviving examples as can be dated to the period after the Norman Conquest that it has suggested the possibility that it is a survival from the Abbey of Edward the Confessor, begun in 1050 and consecrated in 1065. The only known carpenter who worked on the Abbey in 1050 is Teinfrith (Harvey 1974, 41 and 44). This door is illustrated in Fig. 149, wherein the existing part is shown hatched at the edge, and the conjectural restoration is outlined above. This was built of five seasoned oak planks of various widths, cut on the quarter to avoid warp or winding; their edges were rebated as shown in the exploded detail, and the whole was held together by inset ledges – giving a perfectly smooth and plane surface on both sides. The ledges were placed at top and bottom on the front, with a single and central one on the back; their shape is that of two opposed dovetails, the edges forming arcs of great circles. This work was perfectly executed, suggesting that the craftsman possessed very good tools for the purpose – including a router and some type of cutting gauge to cut the insets and rebates accurately to the required depth. Traces survive of the iron

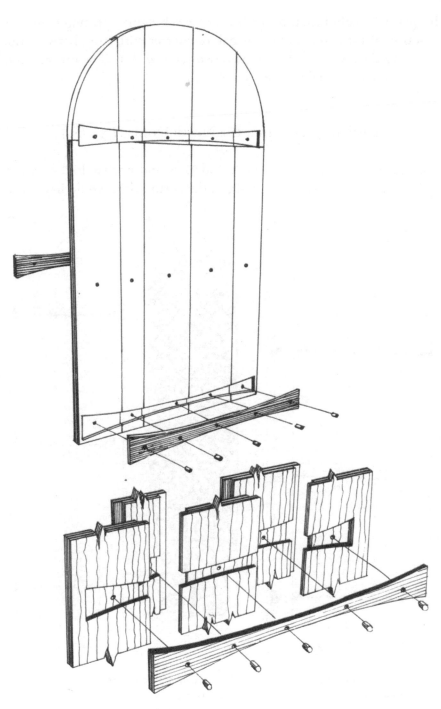

149. Westminster Abbey, the door called 'Pyx'.

hinges originally fitted, in the form of black stains resulting from the reaction of the gallic acid in the oak to contact with iron; betwixt the iron and the timber there was a layer of animal skin. The carpentered assembly was secured by pegs alone.

Durham Cathedral, the North Doors (*Fig. 150*)

The nave was built between 1099 and 1128, and was vaulted between 1128 and 1133 (Harvey 1974, 224), so the north doors would have been

150. Durham Cathedral, the north doors.

151. Durham Cathedral, the south-west doors.

made before 1128. Each leaf was made of five boards, counter-grooved with free tongues, held by three wedge-shaped ledges tapering towards the centre.

Durham Cathedral, the South-West Doors (*Fig. 151*)

The doors now give access to the cloisters, but were made in the same way as the north doors, and later the existing cloisters were added, 1390-1418 (Harvey 1974, 225). Doors constructed in this manner are extremely rare – the only other examples yet known are those of the parish church at Eastwood, Essex (Hewett 1982, 83). The principle was to assemble planks – in this case four – with rebated edges, as shown at lower right of the drawing, and then to drive tapering ledges of semi-circular section into dovetail-sectioned housings routed into their rear faces. These ledges, of course, became increasingly tight as they were driven, and produced what has proved to be a very durable assembly. A vertical section is given at right. The richly-wrought ironwork applied to the face is likely to be of the same age, as it materially strengthens the whole, and a dating contemporary with the nave, 1099-1128 is suggested.

Selby Abbey, the West Doors

The western doors of Selby Abbey are completely different in conception, as is shown in Fig. 152. The doorway itself is considered to date from 1170 at the latest (Kent 1968, 3), and the pair of wooden door-leaves are of the same age. The inside face of the one leaf is illustrated. They are framed in portcullis fashion with squarely crossed ledges on both faces, between which the planking is sandwiched, the whole being roved and clenched through with iron. Inside, the ledges are fitted into the edge-timbers with open notched lap joints, the early form of that joint. Dr. J. Geddes has suggested that 'this type of dense, rectangular portcullis frame with flat washer roves was not fully developed until the fourteenth century' and that these doors may be later replacements (Geddes 1982, 319).

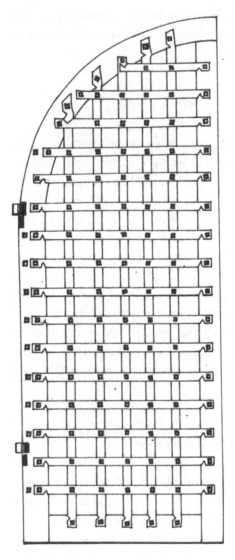

152. Selby Abbey, the west doors.

Peterborough Cathedral, the Central West Doors (*Fig. 153*)

It appears that the great west doors of Peterborough Cathedral were brought up to date when the existing west front was built, between *c.* 1193 and 1230; and that at that time the original planks that had formed the earlier doors were used again. Were this proved, the old planks

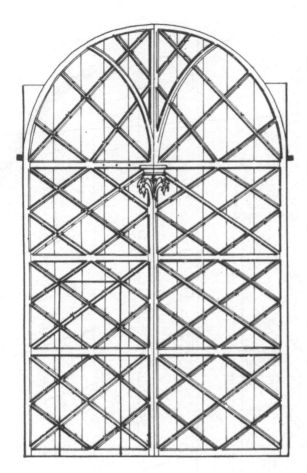

153. Peterborough
Cathedral, the central
west doors.

would probably date from the completion of the nave in 1175 (Harvey
1974, 235). The evidence for their greater age is indefinite, but continued examination reveals the long, 'embattled' edge-joints now known
to have been widely used during the Norman period. The application
of diagonal crossed ledges is possibly an allusion to the dedication of the
great church. The use of dog-tooth ornament, carved on the majority
of the ledges (work which seems not to have been completed) suggests
a date for the alteration close to the cited 1230, as does the carving of
naturalistic foliage on the capital formed by the two shut-timbers. This
represents a springing for two sub-arches of the pointed type then in
vogue – an ingenious way of rendering a Romanesque plan-figure
visually Gothic.

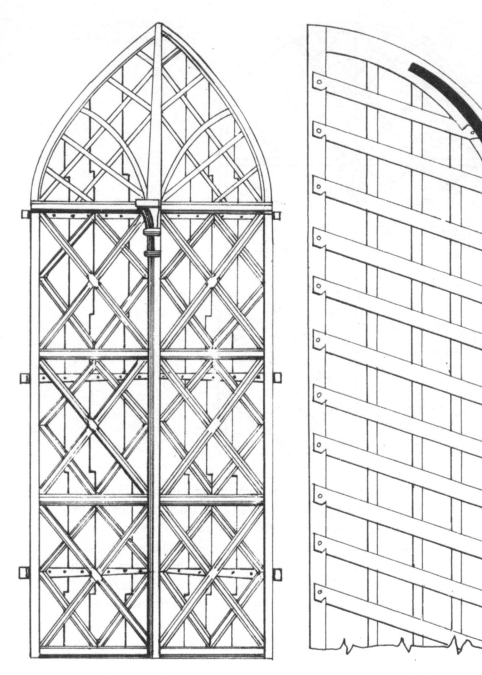

154. Peterborough Cathedral, the
south-west door.

155. Peterborough Cathedral, the west doors of
the precinct.

Peterborough Cathedral, the South-West Door (*Fig. 154*)

The pair of doors in the south-west transept of Peterborough Cathedral also comprised one door originally, made of eight planks, counter-rebated, probably contemporary with the transept itself, *c.*1177-1193 (Harvey 1974, 235) – remade and saltire-braced at a later date, like the central doors of the west front.

Peterborough Cathedral, the West Doors of the Precinct

The gateway to the Peterborough precinct was built during the episcopacy of Bishop Benedict (1177-94), and retains the greater part of its original pair of door-leaves, of which the back of the right-hand half is shown in Fig. 155. These are much worn and have been repaired at the bottom, but retain some early features, among which the method of hanging is rare; the harr-timber was carved to form a pivot and sheathed with iron to resist wear, and rotated inside the iron loop formerly embedded in the masonry – no doubt the lower end was similarly pivoted. As in the Selby specimen the ledges were notch-lapped where fitted into both edge-timbers, and their upward inclination must be interpreted as a measure to prevent sagging.

Ely Cathedral, the Great West Doors

The great west doors of Ely Cathedral, illustrated in Fig. 156, are a case in which re-use is clearly demonstrable. The west end of the nave was reached by 1130, and the west transept and tower were built by 1197 (Harvey 1974, 225). The old planks of the present doors may be of the last date cited. In *c.*1250 the Galilee porch was added and the single Norman door recut and ledged to make the pair of Early English lancet-shaped leaves illustrated. It can be seen that where the two outer curves continued a single-centred arcature could be produced. The horizontal locking-bar shown also dates from *c.*1250. The fronts of both doors were faced by Sir Christopher Wren, whose initials are studded thereon with nail heads.

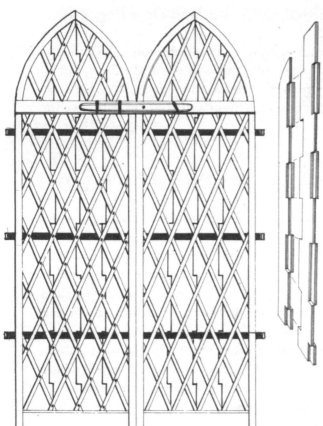

156. Ely Cathedral, the great west doors.

157. Waltham Abbey, Essex.

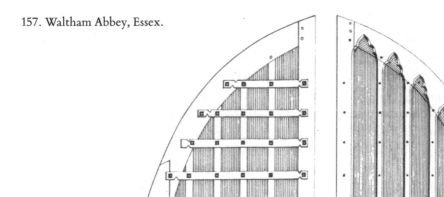

Waltham Abbey, Essex

The head of a right-handed door leaf is preserved at the Abbey as a museum artefact, both faces of which are illustrated in Fig. 157. The harr durn was scarfed, literally, as distinct from jointed; and forms of notched lap joints were used for assembling its portcullis-type rear ledging. The 'shut' was worked into a sally or tace, with a concomitant rebate upon the missing leaf – a rare example of experimentation in this field of carpenters' activities. The cusping is so asymmetrical as to make the period to which it should be ascribed questionable; but the arcature of the durn indicates that the style was pointed and Early English. This leaf probably belonged to the external west portal, which is earlier than the early 14th-century Lady Chapel (Pevsner 1965, 403).

Wells Cathedral, the North Doors

The pair of doors giving access to the nave of Wells from the north porch is illustrated by Fig. 158, which shows the inner and outer face of the same one. Despite their seemingly advanced and elegant design, no reason is known why they should not be contemporary with the doorway and the north porch – built between 1192 and *c*.1230 (Harvey 1974, 241). Within this range they are believed to date more specifically to *c*.1208 (L. S. Colchester, verbal information). The framing is light and excellently cut and chamfered, with slightly heavier hinge-rails, and the two vertical layers of frontal planks are retained by iron roves and clenches. The roves retain the elongated form in which they were originally employed by Scandinavian shipwrights – at least as early as the Viking period. The blind tracery affixed to the fronts, and the band of 'crenellated' ornament at the impost level of the curves, is of later date.

Wells Cathedral, the Door in the Cloister

The door illustrated by Fig. 159 is in the south-east corner of the cloister garth at Wells, and is believed to date from *c*.1200-1230. It has been

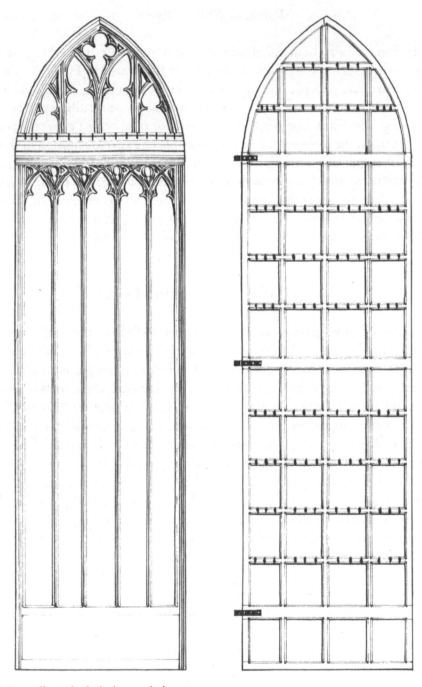

158. Wells Cathedral, the north doors.

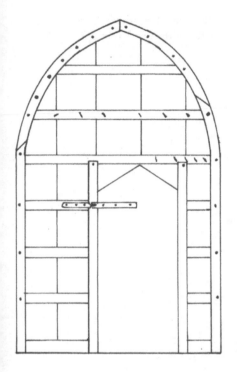

159. Wells Cathedral, the door in the Cloister.

reversed, the original inside now facing outwards, and is made with five planks butted against the frame, which is mortised and nailed.

Lincoln Cathedral, the Door of the North Transept

The back and front of one door-leaf from Lincoln are shown in Fig. 160. The framing is chamfered on every edge and a half-round channel planed along the centre-line of each component – the spikes being driven into this groove. The decorative ironwork is of interest in that it bears no resemblance to characteristically Norman ironwork, yet recalls the anchor-like hinges of the 12th century by the basically C-shaped iron which is bisected by the straight 'ride' strap of the hinges. 'Saltire cross bracing is first found in a datable context in the north transept of Lincoln cathedral 1220-30' (Geddes 1982, 319).

160. Lincoln Cathedral, the door of the north transept.

Wells Cathedral, the West Central Door (*Fig. 161*)

This is believed to date from the construction of the west front, *c.*1230-60 (Harvey 1974, 241). The clenches have squared heads which were carefully set in chiselled counter-sinkings as shown, and closed over roves that were set across the grain of the ledges.

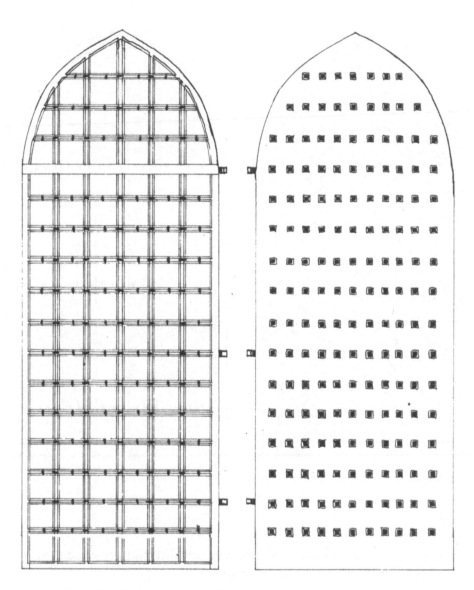

161. Wells Cathedral, the west central door.

Salisbury Cathedral, the West Doors (*Fig. 162*)

The west doors of Salisbury Cathedral are probably the originals, dating from *c.*1266 when the front was finished – under Richard Mason (Harvey 1974, 239). These are of relatively plain carpentry, carefully

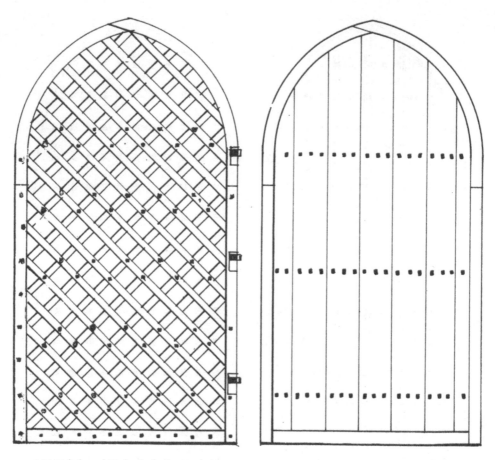

162. Salisbury Cathedral, the west doors.

executed, and have 15 diagonal ledges in each direction, rivetted in vertical alignment through alternate crossings.

Salisbury Cathedral, the Vestry Doors

The vestry was finished between *c.*1260 and 1280. It has three doors as illustrated in Fig. 163, each of three planks with diagonal-bracing.

163. Salisbury
Cathedral, the
Vestry doors.

Wells Cathedral, the Door of the Chapter House Crypt

The door, of which both front and back is shown in Fig. 164, is one of the two ensuring the security of the chapter-house crypt – the treasury – of Wells Cathedral. This undercroft dates from between *c.*1255 and 1286 (Harvey 1974, 241), and the leaves are demonstrably of the same age. The rear framing is similar to that used on the northern doors already shown, while the naturalistic ironwork merits much praise, and would seem to relate to the Early English taste in decorative detail. The fact that the ironwork does not fit the width of the door is not due to a mistake made by the draughtsman – it was made by the blacksmith, who did not allow sufficient plain iron to be covered by the seven-inch returns on either side of the doorway.

164. Wells Cathedral, the door of the Chapter House crypt.

Abbey Dore, Herefordshire, the North Door

This door is built of V-edged planks of oak with three chamfered ledges and a trefoiled head, made of two sections, on its rear face; the whole is fastened with circular pegs of large diameter. The masonry reveals of the aperture have one prominent mould, the roll with frontal fillet, dating from between *c.*1220 and 1260 (Forrester 1972, 31). This door is illustrated in Fig. 165, in which the decorative ironwork is necessarily emphasised, although it is not the proper subject of this work.

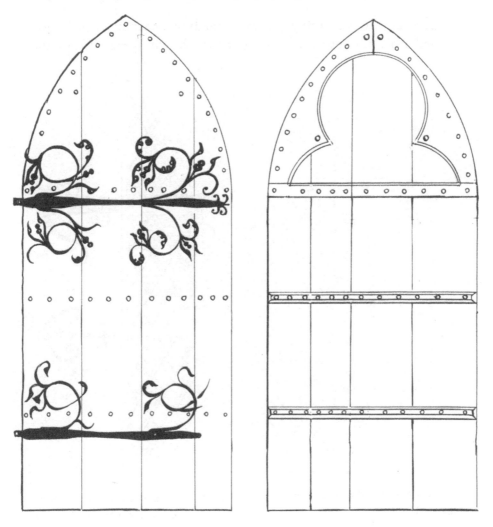

165. Abbey Dore, Herefordshire, the north door.

York Minster, the Door of the Chapter House (*Fig. 166*)

There are two doors into the chapter-house, which was finished by
1286-96 (Harvey 1974, 246). They are made each of seven planks with
fillets diminished on the inside, and ledges diagonally arranged and
halved together, nailed at each crossing.

Ripon Cathedral, the Door of the North Choir Aisle

Among the few surviving pieces of medieval carpentry at Ripon
Cathedral are the severely beautiful doors shown in Fig. 167, giving

166. York Minster, the door of the
Chapter House.

167. Ripon Cathedral, the door of
the north choir aisle.

access from the northern transept into the choir aisle, and probably dating from 1288-97. The front and back of the same leaf are shown, and it is difficult to assess their respective merits — both being visually rich, despite their over-riding restraint.

Wells Cathedral, the Door of the Chapter House

The traceried door with segmentally-arched head shown in Fig. 168 is that giving access to the chapter house from the cathedral, dated to

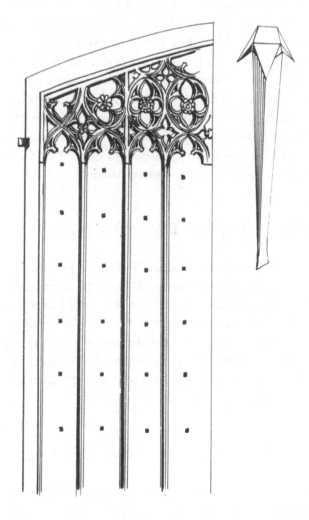

168. Wells Cathedral, the door of the Chapter House.

*c.*1306 (Harvey 1974, 241). Only the decorated front face is illustrated; its rear frame is of squarely disposed and chamfered ledges. It is assembled by spikes, of the type having four points beaten from the corners of their square heads – such as were used in the doors of Salisbury Cathedral (Fig. 162); the four points prevent the head from turning with regard to the surrounding wood.

Westminster Abbey, the Door to the Treasury

The massively-built door-leaf shown in Fig. 169 gives access to the treasury from the cloister of Westminster Abbey. It is dated closely to 1303-7, by reason of a documented burglary which occasioned the construction of the new door (Scott 1863, 290). As drawn, the durns are scarfed and the planking is set between two layers of crossed ledging – all of which is made from heavy, flatly-sectioned oak. The front has squarely crossed ledges with entrant-shouldered barefaced lap dovetails at each end. This last choice of joint is important and highly intelligent, because these ledges were designed in extension, in order that they might resist any tendency of the leaf to sag from its hinges.

Southwell Minster, the North Door of the Nave

The upper half of the front and back of a door illustrated by Fig. 170 represents the north door to the nave of Southwell Minster, an example that is datable by reference to its traceried surface alone. The motif is the reticulated ogee, with quatrefoils inserted in each reticule; the cross sections of the continuous raised muntins conform to rolls with three fillets – which were of maximum popularity between *c.*1270 and *c.*1330 (Forrester 1972, 31). This is, at the earliest, over a century later than the building of the nave, and a date close to *c.*1300 is suggested.

Worcester Cathedral, the Gate of the Edgar Tower

The door leaf illustrated by Fig. 171 survives in the gateway of the Edgar Tower at Worcester, concerning which I am advised that 'the *lower stage*

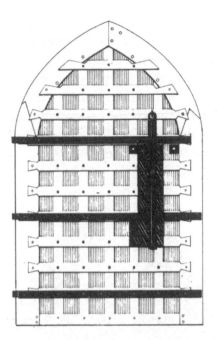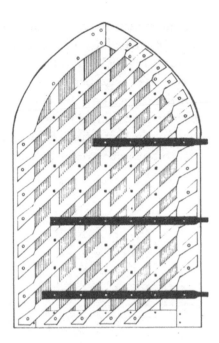

169. Westminster Abbey, the door to the Treasury.

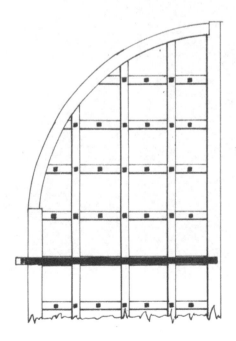

170. Southwell Minster, the north door of the nave.

of the structure was presumably the "new Gate" built in 1346-47 for £41 18s 2d (roll C. 61)' and that the *upper stage* of the tower was building around 1392-93 (Dr. J. H. Harvey, personal communication, 1980). The two leaves together form a single-centred head, which is remarkable for the suggested date. The crossed ledges are on both surfaces with the planks sandwiched between them, a practice similar to that exemplified by the leaves from Selby Abbey and Waltham Abbey, both of early dates. Squared roves and clenches were used at all crossings.

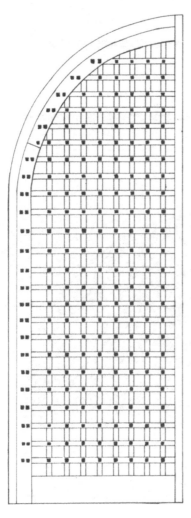

171. Worcester Cathedral, the gate of the Edgar Tower.

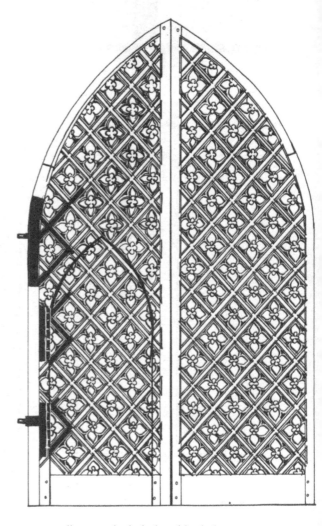

172. St Albans Cathedral, the Abbot's door.

St Alban's Cathedral, the Abbot's Door

Fig. 172 illustrates the cloister door of St Albans Cathedral, known as the Abbot's Door. It is worth stressing that this drawing shows the back or outer face, which is a very fine example of pierced planking with chamfered saltire ledging; it is ascribed to the time of Abbot de la Mare and dated to 1360. The frontal face is decorated with blind and deeply-carved tracery on a larger scale, and the whole is in an excellent state of preservation.

173. Wells Cathedral, the doors of the south aisle screen.

Wells Cathedral, the Doors of the South Aisle Screen

One of the pair of doors in the south aisle screen is shown in Fig. 173. They are believed to date from after 1364 and to be by Thomas Wynford (L. S. Colchester, verbal information). Their central third-height tracery is pierced, while the more complex tracery running into the ogee heads is backed by planking, and blind.

Winchester College, Middle Gate (*Fig. 174*)

'The doors for Middle Gate must have been hung by March 1394 at latest (no accounts survive). So you have dates for doors; Middle Gate, *c*.1393' (Dr. J. H. Harvey, personal communication, 2 September 1980).

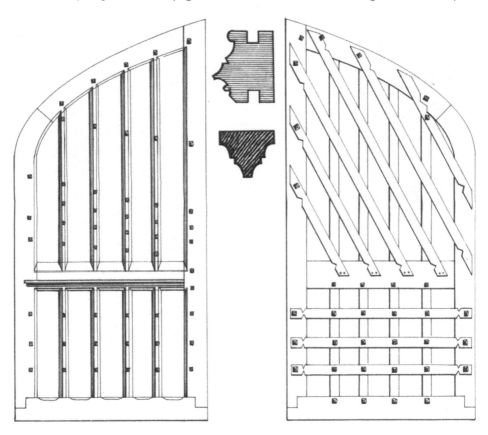

174. Winchester College, Middle Gate.

Winchester College, Outer Gate (*Fig. 175*)

Contracts for building Outer Gate were let to masons and carpenters on 1 November 1394 by William Wynford, master mason to William of Wykeham for New College, Oxford, 1379-1386, and Winchester College; and from 1394, for the nave of Winchester Cathedral, until his death in July 1405. The master carpenter for these works was Hugh Herland, the King's chief carpenter, and overall responsibility for design of the woodwork was his. The gates of Outer Gate tower were hung in the accountancy year October 1396-September 1397 (Dr. J. H. Harvey, personal communication, 2 September 1980).

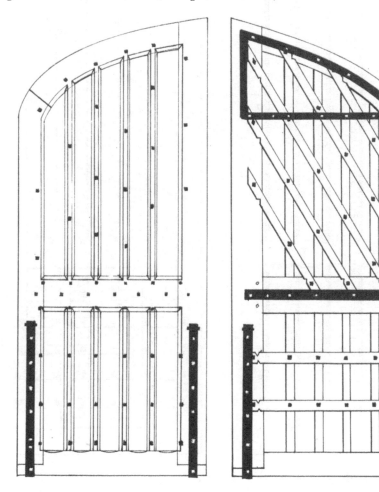

175. Winchester College, Outer Gate.

Winchester, Hospital of St Cross (*Fig. 176*)

The Beaufort Tower was built 1404–47, and the door may date to *c.* 1404 (Dr. J. H. Harvey, personal communication, 2 September 1980).

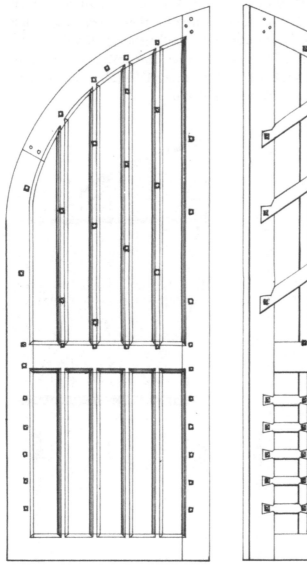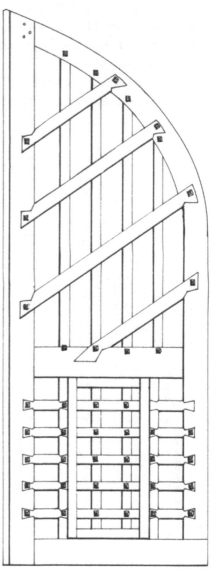

176. Winchester, the gates of the Beaufort Tower.

Canterbury Cathedral, the West Doors

The west doors are constructed in two halves with vertical and central shuts, traceried on the outer face, as illustrated in Fig. 177 which shows the same half-door from both inside and outside. It is possible, in view of the design and quality of workmanship, that these may date from the

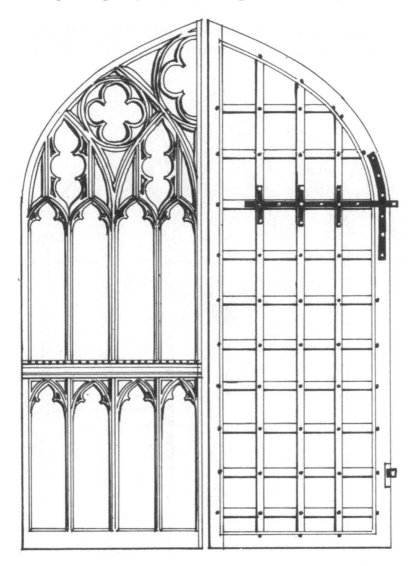

177. Canterbury Cathedral, the west doors.

period of Prior Chillenden, 1391-1411. The crossed ledges of the rear-frames are half-lapped under the stiles and durns, which are chamfered and a quarter-inch proud.

Hereford Cathedral, the Cloister Door

What is now the cloister-door of Hereford Cathedral, illustrated in Fig. 178, probably came from elsewhere, and is of the late 14th century.

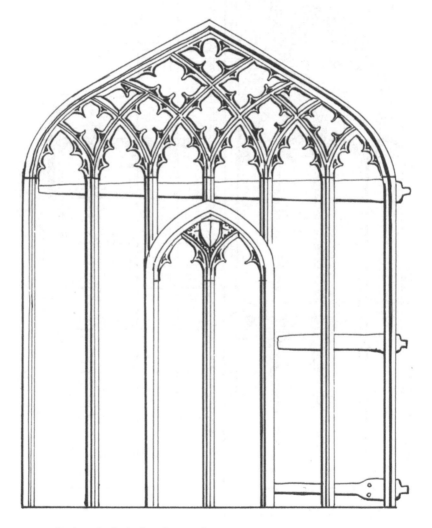

178. Hereford Cathedral, the Cloister door.

Its construction is of crossed planks secured by spikes having re-driven points, to which the tracery was applied.

Gloucester Cathedral, the Doors to the Cloister

The doors giving access to the cloister of Gloucester Cathedral are illustrated in Fig. 179, which gives the back and front of the same leaf.

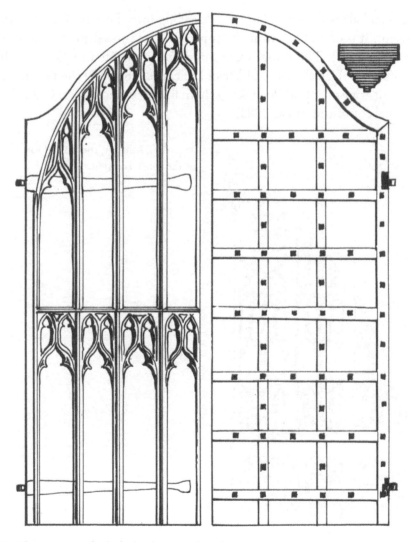

179. Gloucester Cathedral, the doors to the Cloister.

The cloister was designed and built under Robert Lesyngham between *c*.1370 and 1412 (Harvey, 1974, 227). These are painted on their fronts with red and green, having golden fillets. The ironwork is let into trenches, to give a flush and smooth surface, and it is of interest that so Gothic a design should have been shaped to fit a Romanesque aperture.

York Minster, the Main Door of the South Transept

The west half is shown in Fig. 180. 'Leading into the central part of the transept. It is of *c*.1470 and at the top are swords on saltire with a mitre above (St Paul). The doors were lined on the inside in *c*.1740 (in the Gothick style). In the time of Dean Dincance (1858-80) the two leaves were cut across the middle making them like farmdoors' (Dr. E. A. Gee, personal communication, 1983).

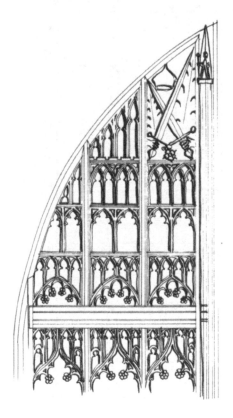

180. York Minster, the main door of the south transept.

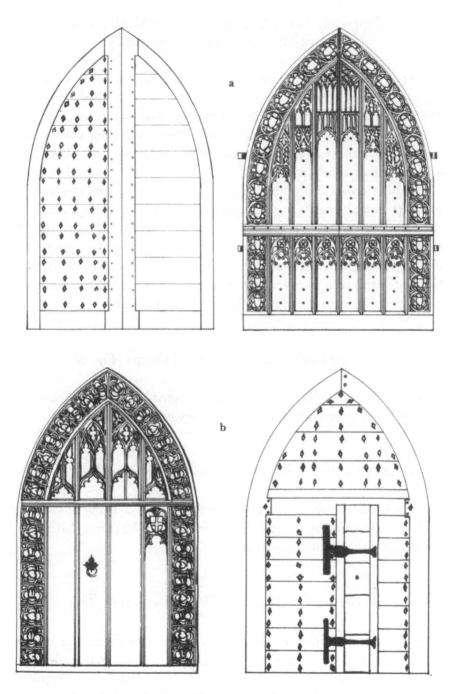

181. Norwich Cathedral, the doors of St Martin's Palace gatehouse.

Norwich Cathedral, the Doors of St Martin's Palace Gatehouse (*Figs. 181a and b*)

Both pairs of doors in the Norwich precinct, which give access to St Martin's Palace, were attributed by J. Adey Repton to Bishop Lyhart, 1446-72, and bear his rebus (Pierce 1965, 34). The larger pair, shown in Fig. 181a, are in my opinion the finest of their kind, since their profusion of rich tracery is wholly successful aesthetically – which was not always true of the later Perpendicular designs. It is noticeable in the majority of doors of the period that the richness of carving lavished on their frontal faces is in clear contrast to the poverty of framing on the rear. This example is formed of heavy plank, in a double laminate within a securely jointed surround, of which the only weakness was due to shrinkage of the inadequately seasoned timber. The doors were finished with lozenge-roves, as previously described, which continued as a decorative form after losing their original purpose.

Norwich Cathedral, the Central Doors (*Fig. 182*)

Made c.1501-36 by Robert Everard, and remade c.1830-6, when the outside was restored (Dr. J. H. Harvey, personal communication).

King's College Chapel, Cambridge, the North Doors

The doors are of interest, the northern pair (Fig. 183) being of the early 16th century, and the western pair dated to 1614-15 (Dr. J. Saltmarsh, personal communication, 4 January 1973).

Westminster Abbey, the West Doors (*Fig. 184*)

The existing west doors of Westminster Abbey have a secular appearance despite their conformity to the doorway. I am given to understand that they were made a little before 1700 (S. E. Dykes-Bower, Architect

182. Norwich Cathedral, the central doors.

183. King's College Chapel, Cambridge, the north doors.

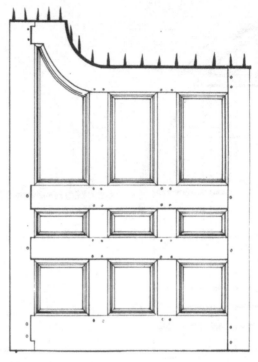

184. (*left*) Westminster Abbey, the west doors.

185. (*above*) Wells Cathedral, gates of the north porch.

to the Abbey, verbal information). They were built with fielded panels set into the grooved edges of the complex square framing, of excellent workmanship; their shutting edges have complicated cross-sections.

Wells Cathedral, the North Porch (*Fig. 185*)

*c.*1750.

HOISTING MACHINERY

The lifting and delivery to various parts of the site of the enormous quantities of dressed stone, or timber, used in building a cathedral postulates the use of mechanical aids. Consider, for example, the raising of the 5,000 tons of Chilmark stone to build the spire at Salisbury, which *begins* at a height approaching 300 feet! Reference to such 'engines' are numerous in contemporary accounts. Salzman quotes the Westminster accounts for 1532: 'John Moulton of Westminster, mason, for two paire lowettis of iron by him delyverd for the saufe cranage of stone redy wrought from breaking oute of shippes and lighters' and also: 'To the same for two payre of doble robenettes with their poleys of brasse and other apparelle … for hoysing of wrought stone'. Salzman, 'with some hesitation', suggests that 'robenettes' were 'grips' or 'scissors' – by which the stones were individually attached to the lifting rope. More pertinently he quotes the accounts of King's College Chapel for 1507, wherein 20d was paid to 'Richard Symson for caryage of a cheste wyth gynnes and robynettes for the cranes, weying ij cweyght, from London to Cambridge'. And earlier, with reference to Winchester in 1222, he cites tallow purchased for 'greasing the machines (ingenia) for lifting timber and water'. The principle of all such devices was a rotary windlass barrel, from which the lifting rope could be led through suitable sheaves – the 'poleis' (pulleys) of the ancient records – to various points about the site, possibly to a 'faulcon' or jib. The barrel which wound or unwound the rope was turned by means of a large wheel, the whole being known as a 'verne' (Salzman 1952, 323-4).

The surviving examples are described and illustrated in what seems to be their chronological order, and some comparisons made thereafter.

The Windlass at Tewkesbury Abbey

Tewkesbury Abbey was a Benedictine foundation of the end of the 11th century, extended during the 12th; the lower part of the tower was completed by 1121, and the top was added about 40 years later (Maclagan 1971, 3). It is 46 ft. square and 132 ft. to the base course of the existing battlements. A long-disused windlass with runged felloes is on the floor beneath the bell chamber, illustrated in Fig. 186. It has a rim diameter of 12 ft., built with six compass-arms that transfix the shaft and are wedged in place. The six felloes are simply scarfed, and these joints are each impaled by one rung. The iron gudgeons remain at either end; the rope-drum is larger in diameter than the shaft itself. (I am indebted to Mr. J. K. Major for information about this example.)

The Windlass at Durham Cathedral

The Durham windlass is now fragmentary, standing in mountings set into the reveals of a window in the north-west tower. To both the west towers two building dates apply – 1128 for their bases, and *c.*1220 for their upper stages (Harvey 1972, 129). The windlass illustrated in Fig. 187 is strange in possessing not only one squared end, pierced for two compass-arms, but also two mortises for a similar pair of arms on its rounded length. Whether this is an item of wind- or watermill machinery cannot be determined. The diameters of these two wheels would have been less than is normal in cathedral windlass wheels, but the probability is that this was one – due to its survival in this situation. It is difficult to envisage a great cathedral using second-hand items to build a make-shift machine for its own use.

186. The windlass at
Tewkesbury Abbey.

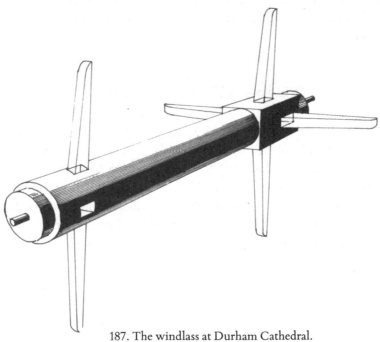

187. The windlass at Durham Cathedral.

The Windlass at Peterborough Cathedral

The Peterborough windlass is complete and in working order. It is illustrated in Fig. 188, without its mountings, which are framed in timber and now set into a western nave triforium space. This is a very large wheel, over 12 ft. in diameter, with its four compass-arms nicely chamfered and its four felloes closely scarfed together. The rungs are long enough, and stout enough for a man to 'climb' either side of the rim. The lifting power of this, given two 10-stone men on the rim and the purchase of over six feet of spoke, must have been enough to lift any building stone.

188. The windlass at Peterborough Cathedral.

The Windlass at Salisbury Cathedral

The Salisbury windlass illustrated in Fig. 189 is still in regular use. It is mounted inside the lowest stage of the spire-scaffold, and appears to be of a design derived from the compass-arm type, having a hub-like 'box' built around its shaft at the spoke point. This cannot be examined, but two of the arms may well pass through the shaft, and the others rely on

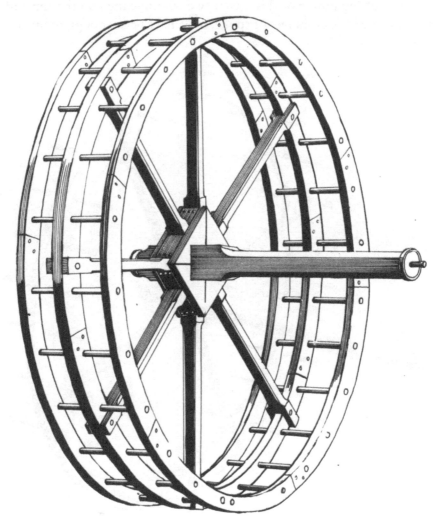

189. The windlass at Salisbury Cathedral.

the boxing. The triple ring of felloes with the staves driven through provides a very strong rim, either for walking inside or for climbing outside – the diameter is large enough for either system. It is probable that this great wheel dates from the first building of the cathedral, *c.*1220, because such a device was essential during that operation, and has never been so necessary since, while its sheer quality argues against it being a subsequent replacement.

The 'Faucon' at Canterbury Cathedral

Two further items belonging to the hoisting aspect of cathedral building survive, and both may be what early records term 'faucons' – as at Winchester in 1257 (Salzman 1952, 324) – or elsewhere as 'hawks'. In modern terminology these are jibs, and serve to hold the lifting rope clear of walls and other obstructions. They were wall-mounted and could rotate through 180 degrees in the horizontal plane; their load could thus be deposited at any point along the radius of their arms. The one illustrated in Fig. 190 is in the Bell Harry Tower at Canterbury. It seems to relate specifically to the great wooden boss in the centre of the vault, since the strength of its construction would hardly allow of raising heavier items than that boss. Therefore it is probably datable to the building of the vault by John Wastell in *c.*1505 (Harvey 1974, 221), and its details accord very well with that ascription. It is mounted inside the tower in the south-eastern corner, and still has its rope, attached to a small hand-spiked windlass.

The 'Faucon' at King's College, Cambridge

The other example, illustrated in Fig. 191, lies in the northern external passageway of King's College Chapel, at the level of the vaults. The ironwork is attached by forelock-bolts, and it has brass sheaves at either side of the arm with guide-staples for the rope – all designed in a manner reconcilable with the 16th century. The machine to which this was an essential adjunct has not survived. It may relate to Richard Russell's work there, *c.*1510.

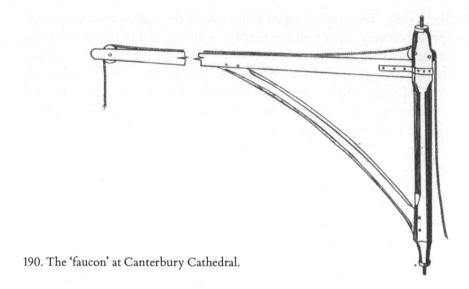

190. The 'faucon' at Canterbury Cathedral.

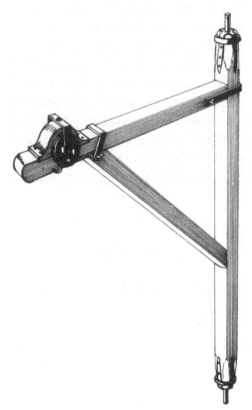

191. The 'faucon' at King's College, Cambridge.

The Treadwheel Windlass at Beverley Minster

The tread-wheel windlass shown in Fig. 192 is mounted above the central crossing of Beverley Minster. This is built to the clasp-arm design, and has two peripheral felloe-rims that are literally 'shrouded' to provide a walking surface within the wheel, which is more than large enough for that purpose. As in the Salisbury wheel a need for secondary arms was here recognised, and these were fitted by halving them through braces betwixt the primary arms. According to my site-notes the iron straps fixed to the arms' crossings look newer than the woodwork, and they may be later repairs.

The Treadwheel Windlass at Canterbury Cathedral

A very large tread-wheel is mounted within the upper stage of the Bell Harry tower at Canterbury, illustrated in Fig. 193. This is of clasp-arm construction, and considered at the time of inspection to be similar in date to the Beverley wheel.

The Windlass at Norwich Cathedral

The 'helm'-type wheel and windlass illustrated in Fig. 194 is situated in the base of the spire of Norwich Cathedral. It was certainly in that position when J. Adey Repton's plate of the spire was engraved at the end of the 18th century, and it is almost certainly older. This is over 12 ft. in diameter: the spokes were hauled down to operate the barrel, which has 'kelps' spiked on to give the rope a grip, the whole item having a strongly maritime character. The use of screw-threaded bolts for this assembly dates it as later than the others, but it could be of the close of the 17th century. The construction depends upon two discs of elm which are mounted on the squared part of the shaft, and the spokes are trapped between these and bolted; whilst the two thin rims of felloes act as distance-pieces between the spokes, preventing their breakage.

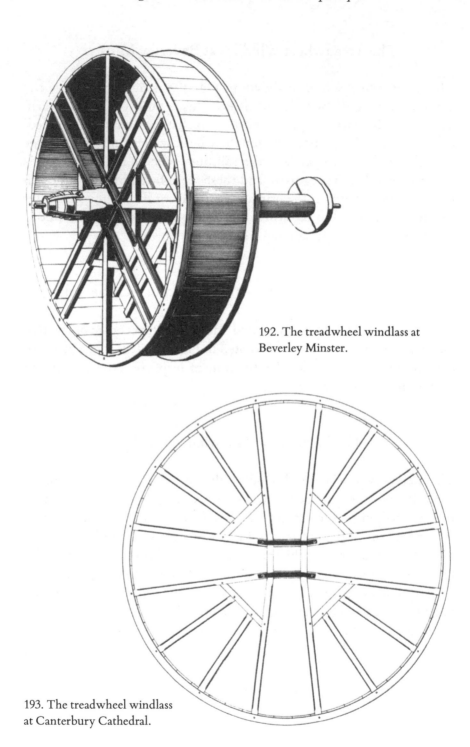

192. The treadwheel windlass at Beverley Minster.

193. The treadwheel windlass at Canterbury Cathedral.

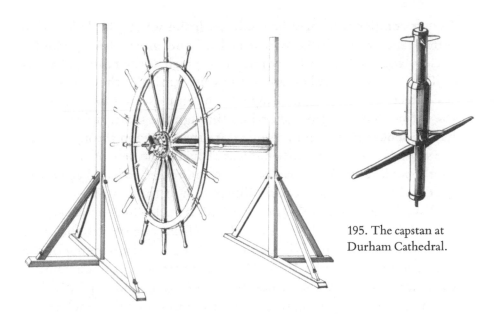

195. The capstan at
Durham Cathedral.

194. The windlass at Norwich Cathedral.

The Capstan at Durham Cathedral

Another item having distinctly maritime associations is the vertical capstan shown in Fig. 195. This is mounted in the south-western tower of
Durham, and being mainly of softwood is considered to date from the
18th century.

The Geared Windlass at St Paul's Cathedral (*not illustrated*)

St Paul's Cathedral possesses a geared windlass with reduction drive,
believed to date from Wren's operations, mounted in the Golden
Gallery beneath the Ball; and also a vertical oaken capstan built into the
north-west bell turret.

Little has been published concerning the historical development
of hoisting wheels, although a great deal of information is available.

Direct comparisons may be made with the wheels built for both wind- and water-mills, whilst a third and more closely-similar variety was used for military purposes – such as the raising of portcullises. The structural design of the wheels in all of these categories was affected by one major change, from compass-arms to clasp-arms, the former having spokes mortised through the shaft, and the latter having spokes framed into a square or other polygon through which the shaft passed, and by which it was clasped. The date of this transition is not known, but it has long been considered to have been about 1556, when the earliest known illustration of the clasp-arm type of mill appeared (Singer 1956, Fig. 19). This is misleading, however, and all that can be inferred is that the change from one type to the other was slow, and occupied the end of the 17th and the opening of the 18th centuries (Hewett 1968, 70-74). A great deal of documentary evidence contradicts the earlier view. The sketchbook of Villard de Honnecourt, comprising some personal notes of that French architect, who is believed to have been working between *c*.1225 and *c*.1250, illustrates an example of the clasp-arm wheel (Bowie 1959, 130). This is reproduced in Fig. 196, and de Honnecourt wrote on his drawing that it showed how to 'brace the spokes of a wheel without cutting the shaft'. His sketch is most ingenious, and actually uses the same principle of clasping-arms twice, by means of cross-halvings and lap-dovetails. Perhaps during the same period, and certainly during the early 13th century, the Round Table now exhibited in the Hall of Winchester Castle was built, and this also used the clasp-arm principle – despite the fact that it was not a wheel. This is illustrated in Fig. 197. One other technological development is also involved, and once more its date of introduction is not known. This was the safety mechanism, a ratchet-wheel with pawl, that prevents the rotation of the windlass barrel in the opposite direction to that intended. This principle was certainly known as early as *c*.1386, having been used for the barrels of the Salisbury Cathedral clock (Backinsell 1977, H.M.2), but it is as yet unknown in the context of cathedral windlasses, wherein the great wheels rotated freely in either direction. One windlass is known that does incorporate pawls and ratchet-wheels, and this is, additionally, a two-shaft mechanism that provides 'reduction gearing' without recourse to cogged wheels. It is

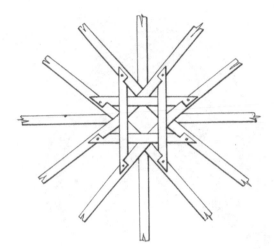

196. A clasp-arm wheel illustrated by Villard de Honnecourt.

197. The sub-structure of the round table at Winchester Castle.

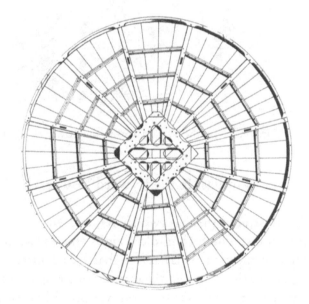

illustrated in Fig. 198, and survives in working order in the Byward Tower, Tower of London – the fabric of which is ascribed to 1260. It is possible, in the light of all the evidence, to ascribe the surviving cathedral specimens to approximate dates, after the two phases of construction discernible in the Byward Tower windlass have been assessed in that respect.

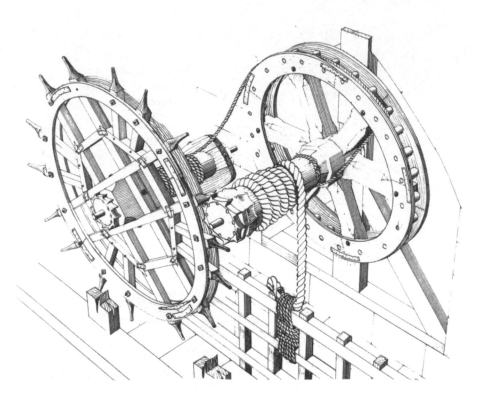

198. The windlass in the Byward Tower, the Tower of London.

The Windlass in the Byward Tower, the Tower of London

As illustrated this has two shafts, and was used to raise or lower the portcullis, which I am told weighs almost two tons, and can be operated by one man. This has two shafts, each fitted with a ring of ratchets spiked to their ends, in the manner of a ring of wedges, driven betwixt the iron and the oak; the iron was possibly grooved to accommodate these on its internal circumference. The six arms, or spokes of the rear wheel are radial and tenoned into the shaft, whilst its double rim of felloes form a 'lantern' pinion round which the secondary rope is coiled. The chamfering of both shaft and spokes would be acceptable as of the 13th century – but could equally well be later. Only forelock-bolts appear on this wheel. The action of the machine was to raise the portcullis by winding the heavy rope around the first barrel, which was

achieved by unwinding the lighter rope from off the lantern-pinion
– winding its other end onto the second barrel, which was turned by
pulling on the projecting spokes. This second wheel, rather like that
of the Norwich Cathedral machine, which also has projecting spokes,
was clasp-armed and assembled with screw-threaded nuts and bolts.
Whilst it cannot be proved that these screw-threads are original, it is
difficult to envisage such a wheel being strong enough without them
– and previous forelock-bolts would have left some evidence of their
removal. The two shafts would, therefore, appear to date from before,
and after, the important change from compass- to clasp-arm construction. The monumental authority of Salzman once more assists with
dating these events, for from the Westminster accounts of 1532 he cites
a payment for 'skrewis, and vices to the same, provided for a dore': he
found no earlier reference to threaded bolts and considered their introduction to date from then, *c.*1532 (Salzman 1952, 309). The Tower
being the capital stronghold of the monarch, we are safe to assume
that such developments were there exploited almost as soon as known.
The hand wheel could therefore date from the later reign of Henry
VIII, and the lantern wheel from the initial building and fitting-out
of the Byward Tower, *c.*1260. This provides two classifications for the
cathedral wheels, compass- or clasp-arm, with the second category
sub-divided by the use of forelock- or threaded-bolts. A reappraisal
of the items requires to be made with regard to the *style* of carpenters' work and chamfering, and the totality of comparable examples in
mind, together with the early examples from Villard de Honnecourt
and Winchester Castle.

In general these 'vernes' or 'great wheels' would have been constructed as soon as they were required, or the building attained any
height; thereafter, as the early records clearly show, they were maintained in working order for as long as they were needed. Replacement
with new machinery was unlikely, and only major restorations can assist
with the dating of more recent windlasses. The compassarm wheels exist
at Tewkesbury, Durham, Peterborough and Salisbury, probably in that
order chronologically. Salisbury has an advanced and superior form of
the type, which can be ascribed to the general building of that cathedral
between 1220 and 1266, after which it would have been re-sited for the
building of the spire. The earlier and simpler wheels are probably those

provided for the various building operations, the dates of which have been given. The two clasp-arm wheels, of Beverley and Canterbury, differ widely in their workmanship. The Beverley example may date from the activities of Nicholas Hawksmoor, which began in 1716; and the Bell Harry wheel may be even later. The Norwich wheel has screw-threads that are necessarily original, and it must date from somewhere between their introduction and the Repton engraving, which shows it in position at the end of the 18th century. The Durham capstan is mainly of softwood, and of the 18th century, whilst the two jibs, or falcons, seem to relate closely to the given dates of their buildings.

JOINTS

Lap Joints

Plate 1.

Rayleigh Castle, Essex, was excavated in 1909 by Mr. E. B. Francis, who read a paper describing his findings at a meeting of the Essex Archaeological Society in 1910. The large hollow oak post which was found is unique in England, in that the diagonal brace recessed into it has never been seen in that form before or since. This early fortress is one of the few strongholds mentioned in Domesday. Mr. St John Hope stated that it was one of a ring of castles built under the Conqueror's orders, and he had no doubt that the earthworks at Rayleigh were made by Suen, son of Robert Fitz Wimarc. If this is the case, the Castle must have been built sometime between 1066 and 1087 (Francis 1913, 147-9).

'Castles were by far the most important instrument which the Normans used for the subjugation of the country. Before the Conquest they had been almost unknown in England, for they were basically different from the fortified towns or boroughs of the Anglo-Saxons. The boroughs were large because they had been designed as places where all men could take refuge against external invaders. The castle, on the other hand, was small and was designed as a place from which a few men could dominate a subject population. Basically it was a wooden tower erected on a mound of earth (or *motte*), the purpose of which was to defend the base of the tower from incendiaries, while also being steep enough to make a cavalry-charge impossible. Though small in comparison to a borough, the building of a *motte* must have been a major undertaking' (Davis 1966, 284).

Fig. 199

From the Rhenish helm of the church of St Mary, Sompting, Sussex, where it is used for the wall anchors. These joints were withdrawable, and were designed only to resist shearing stresses exerted in the plane of the roughly triangular figure which the assembly produced. Their date is uncertain, but must lie between *c.*950 and *c.*1066 (Hewett 1980, 20).

Fig. 200

Another form of lap joint, found in a re-used context, upon one joist of the floor in the Sompting tower (Hewett 1980, Fig. 18). This diagram shows the joint as a plan view; both its date and its original purpose are uncertain.

Fig. 199

Fig. 200

Fig. 201

From the barn of Paul's Hall, Belchamp St Paul, Essex. The most interesting feature surviving from the previous use is the lap-joint matrix, now empty but retaining a polygonal peg, on the major stud. This is a form of the stopped and curved lap-joint (Hewett 1980, 23-4).

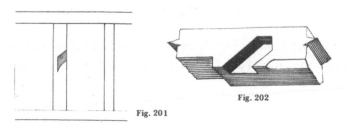

Fig. 201

Fig. 202

Fig. 202

From Moyses Hall, Bury St Edmunds, Suffolk. This is of the late 12th century, the earliest domestic building in Bury. It is of stone, of two storeys with a 19th-century east wall, and probably connected with the Abbey. It is said to have been built by a rich Jew (*cf.* 'The Jew's House' at Lincoln). The joint is defined as a stub-mortise with two abutments.

Fig. 203

From the Norman Hall, Peterborough, *c.*1155-1177. There are four different profiles, as shown, perhaps the earliest open notched lap joints of archaic entry-angle seen in England – difficult to date, but possibly earlier than those in the re-used roof timbers of Waltham Abbey.

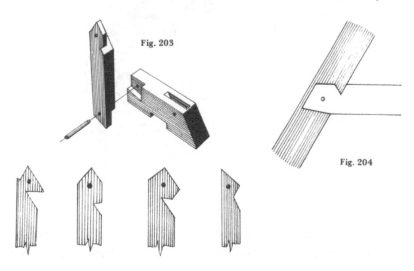

Fig. 203

Fig. 204

Fig. 204

From Waltham Abbey. This is an open notched lap of archaic profile, not very strong; it had a 'nick' without the refined entry-angle developed later.

Fig. 205

From the nave roof of Peterborough Cathedral, a fragment on the north side, visible from the west front. The nave was begun *c.*1155 and finished 1175 (Harvey 1974, 235). This is a better open notched lap, but it is still of archaic profile.

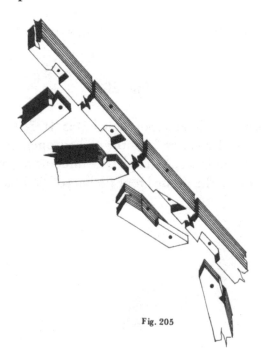

Fig. 205

Fig. 206

From a fragment of re-used timber at Abbey Dore, Herefordshire, founded in 1147; it is tending towards the refined profile, but not yet good enough.

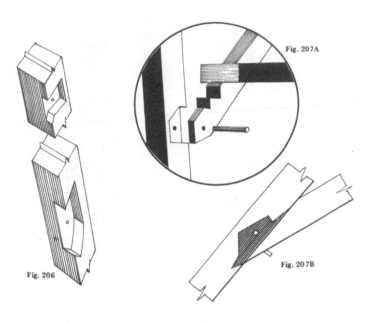

Fig. 207a

From the lean-to roof of the north triforium of Wells Cathedral, *c*.1175-1192. This is an open notched lap of archaic profile.

Fig. 207b

An open notched lap joint from the lean-to roof of the south triforium of the choir of Wells Cathedral. This has a refined entry combined with an extended 'tip' of uncertain purpose. Visually it resembles the socket existing on the re-used joist of the Sompting tower floor, and being situated in the eastern arm of the church – a part which was built between *c*.1176 and 1184 – it may date from *c*.1180.

Fig. 208

From the roof of St Hugh's Choir, Lincoln Cathedral, *c*.1192-1200. This is of archaic profile, made very well and long, but is still called an open notched lap joint.

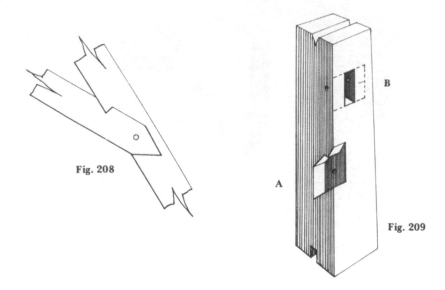

Fig. 208

Fig. 209

Plate 3

From Navestock church, Essex. The present building dates from the
11th century, when it undoubtedly consisted only of a nave 44 ft. by
20 ft. and a sanctuary. The splendid timber-framed belfry was added
to the south-west of the nave sometime before 1250, originally free-
standing, and about that date the south aisle and chapel were added,
connecting with the belfry, and the chancel was probably rebuilt. The
belfry has one notched lap of archaic profile, which has broken under
extending stress at an early date; the 'refined' form of it was stronger
(Hewett and Smith 1972, 82-5; Hewett 1982, 62).

Figs. 209a and b

From the barn, Grange Farm, Coggeshall, Essex. It is situated half
a mile from its proprietor, the Savignac abbey of Coggeshall, a royal
foundation of *c.*1147. The proximity suggests a slightly remote *curia* of
the main settlement rather than a typical detached Cistercian grange,
and 1140 is a valid *terminus post quem* for its building. The foot of one
north-eastern post has been carbon-dated to 1020±90 (Harwell 1976,

HAR-1258) This is the earliest example known with 'refined' angle of entry, offering maximum objection to withdrawal. The mortise has a low aspect ratio when viewed on plan; the length is little greater than the width (Hewett 1980, 47-50).

Fig. 210

From Canterbury Cathedral, the south-east spirelet, *c.*1174. The secret notched laps are of archaic profile. Three others of similar type and date to this are: The Old Court House, Limpsfield, Surrey (Fig. 271); a manor house in Harlow Essex, called Harlowbury; and Cogan House, Canterbury (Parkin 1970, 123-137).

Fig. 211

From the Barley Barn, Cressing Temple, Essex, carbon-dated to 1200±60 years. These are open notched laps with refined profiles, without any diminution in depth (Hewett 1980, 59-63).

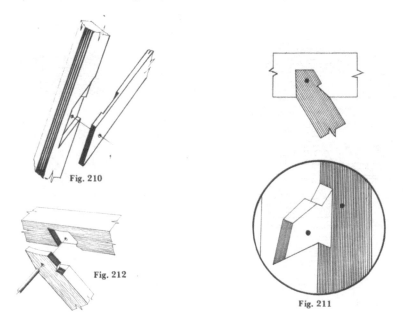

Fig. 210

Fig. 212

Fig. 211

Fig. 212

From Wells Cathedral, the eastern part of the nave, between *c.*1180 and *c.*1209, before the 'break' of the Interdict. These are open notched laps of refined profile.

Fig. 213

From Worcester Cathedral, in re-used timber in the roof of the nave, and probably contemporary with the two western bays added *c.*1175 (Harvey 1974, 244). The open notched laps are of refined profile.

Fig. 214

From Lincoln Cathedral, the roof of the Consistory Court and Bellringers' Chapel, *c.*1225-35. These are open notched laps of refined profile.

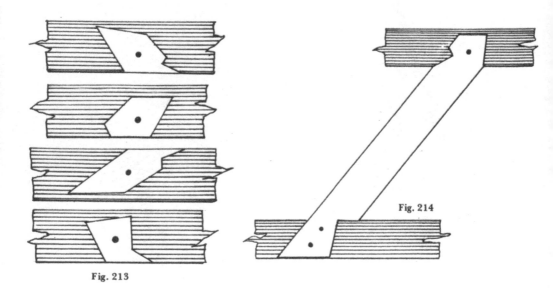

Fig. 214

Fig. 213

Fig. 215

From Wells Cathedral. The roof of the nave was continued westwards after the Interdict, to its termination at the front during the life of Adam Lock, the architect. The transverse design was unaltered, and the rafter couples immediately west of the point where building had ceased in 1209 are datable to 1213 and later. The difference between the two lengths of roof was only that of joint type, the 'secret' form of notched lap joint being used after the 'break'. These are of refined profile but not yet diminished in depth.

Fig. 216

From Salisbury Cathedral, the north-east transept, *c*.1220 to *c*.1237. This is a notched lap of 'secret' form, but not of refined profile, without diminution in depth.

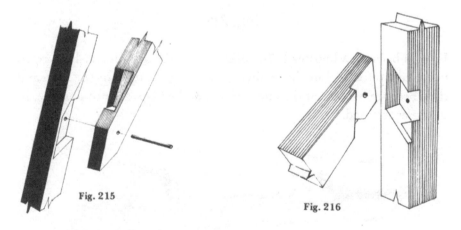

Fig. 215

Fig. 216

Fig. 217

From Romsey Abbey, *c*.1230. The soulaces are illustrated at *a*, open notched laps of refined profile, and the collar beams at *b* used secret notched laps with archaic profile. Also illustrated in Plate 4.

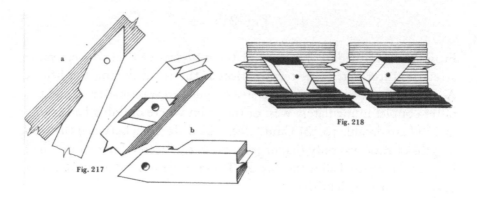

Fig. 218

From the Wheat Barn, Cressing Temple, Essex, carbon-dated 1255±60 years. Both joints are on the same timber, without refined entry, one being open and the other secret (Hewett 1980, 102-5).

Fig. 219

From a barn at Whepstead, Suffolk. The tie-beams mostly have open notched lap joints, but the timber is not long enough for the span and they become 'waney' at the ends. Here notched laps are not possible, so mortises are used.

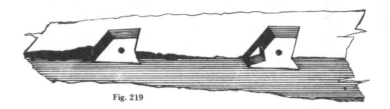

Fig. 220

From Westminster Abbey, the eastern apse, by Master Alexander, the best medieval architect in England of the time, and Master Carpenter to the King, 1245 to 1259. This is a notched lap joint with refined profile and diminished lap, spurred behind, with two face-pegs. It is the best notched lap joint yet seen.

Fig. 221

On re-used rafters of a barn at Parkbury Farm, Radlett, Hertfordshire, believed to have come originally from the nave roof of St Albans Abbey. 'The earlier work at Parkbury can be equated with the change of policy at St Albans in 1195 when it was decided to extend the west front by a further three bays. Very little, though, happened until 1214-35 when William of Trumpington speeded the work on to completion. If the Parkbury rafters came from the church then this would be the vital period' (Gibson 1976, 161-2). The notched laps are of refined profile, and also have diminution in the depth.

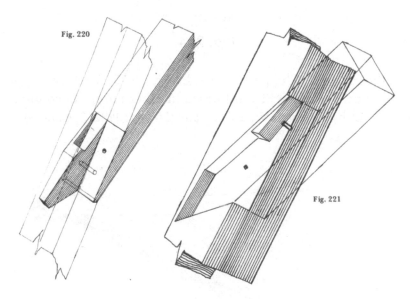

Fig. 220

Fig. 221

Fig. 222

From the south transeptal roof of Exeter Cathedral, between *c.*1190 and *c.*1310. This can be defined as a counter-sallied cross-halving.

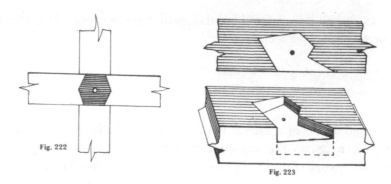

Fig. 222

Fig. 223

Fig. 223

From Wells Cathedral, the south choir transept, dedicated 1330. This is an open notched lap of refined entry, diminished in depth towards the notch.

Fig. 224

From the church of Chipping Ongar, Essex, *c.*1075. A different type of lapped joint, and one which is known in other European countries – until late 12th century in West Germany (Hewett 1982, 3-4 and 8).

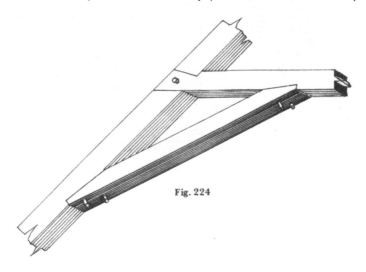

Fig. 224

Other Carpentry Joints

Fig. 225

From the church of Bradwell-juxta-Coggeshall, Essex. A single and central tenon mounted upon spurred shoulders, between *c*.1080 and *c*.1130 (Hewett 1980, 52-3).

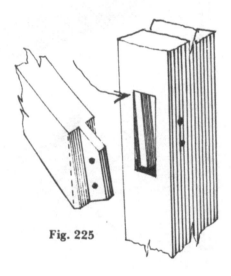

Fig. 225

Fig. 226

From Ely Cathedral, the great west doors. The evidence for re-use lies in the assembly method of the eight planks by means of counter-rebated edges, shown in perspective.

Fig. 227

From Lincoln Cathedral, the north triforium of the Choir, 1192-1210. This scarf is splayed and tabled with three face-pegs. Also used *c*.1336 at Bisham Abbey, Berkshire, with four pegs.

Fig. 228

From Salisbury Cathedral, the north choir triforium, 1225-37. This scarf is bridled with through-splayed shoulders, *c.*1237-40. By the architects Nicholas of Ely (and Elias of Dereham).

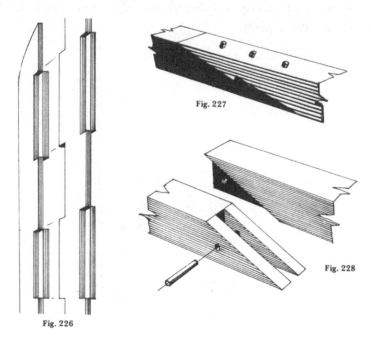

Fig. 227

Fig. 228

Fig. 226

Fig. 229

From the Greyfriars Church at Lincoln, begun *c.*1237. The curved timbers of the archivolt are fitted with diminished square tenons, with an angular fixture.

Fig. 230

From Lincoln Cathedral, the Angel Choir. This scarf is stop-splayed with under-squinted and sallied abutments, tenoned and pegged, with two face-pegs. By Simon of Thirsk, sometime between 1256 and 1280.

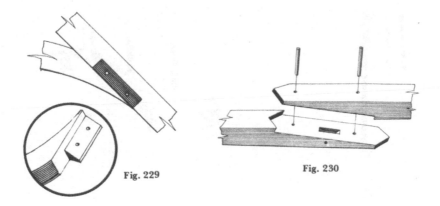

Fig. 229

Fig. 230

Fig. 231

A queen-post also from the roof of the Angel Choir. It has a tenon at the top and a dovetail at the bottom, and two doubled iron straps with fore-locks.

Fig. 232a

From Salisbury Cathedral, the great transept, 1237-58. Front view of the double wall-plates in the triforium, with arcaded front and decorative pendants.

Fig. 232b

Rear view of the same assembly showing the in-canted ashlars and the method of tenoning the sole-pieces into the return of the double plates.

Fig. 233

From Salisbury Cathedral, the spire, 1320-87. Splayed and tabled scarf with bridled upper abutment, edge peg, and face spike. This joint was used in the lower framing of the spire scaffold. The causes of its

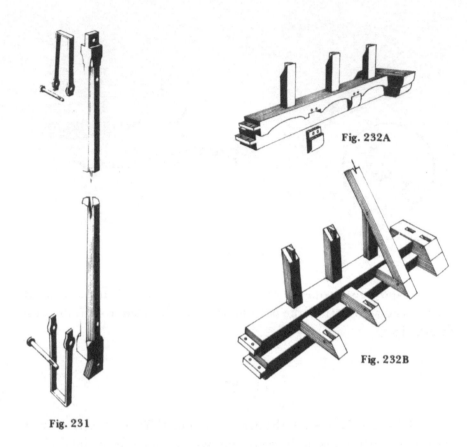

Fig. 232A

Fig. 232B

Fig. 231

invention are at the time of writing obscure, and no further examples of its use are known. It was, however, influential, because it combined for the first time the principles of bridling, tabling and splaying of scarfs.

Fig. 234

From the Old Deanery, Salisbury, 1258-74. This building incorporates what may be the earliest example of splayed scarfs with tonguing. It unites the lengths of top-plate and may be defined as: stop-splayed with tabling and square under-squinted abutments, tongued and grooved part-length, laterally keyed and with one edge-peg.

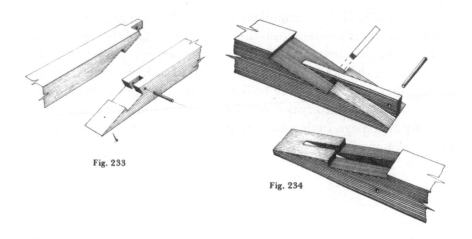

Fig. 233

Fig. 234

Fig. 235

From St Mary's Hospital, Chichester, Sussex. The earliest reference to this hospital is dated 1229, when a charter of Henry III specifically referred to it as existing on a corner site at the junction of South and East Streets, Chichester. It stands today in St Martin's Square, a site that was vacated by the Grey Friars Minor in 1269, and in 1272 Edward I confirmed the hospital on its new site, with exemption from the Statute of Mortmain. The main infirmary and chapel were completed about

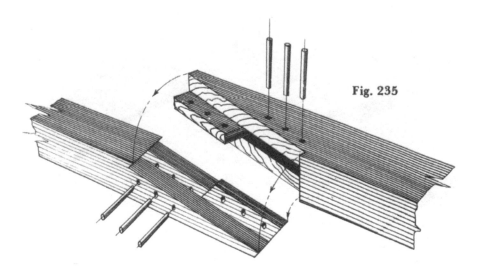

Fig. 235

1290 (Powell 1974, 1). It is unlikely that the Hall housing the infirmary was ever physically moved from the earlier site, since if that were proved it would incorporate a scarf far in advance of any surviving in cathedral carpentry of the time. The drawing illustrates a hypothetical internal arrangement. The joint may be defined as: stop-splayed with square under-squinted abutments, the splays counter-tongued and grooved, with six edge-pegs. In this building these scarfs are used in unsupported positions along the top-plates.

Fig. 236

From Salisbury Cathedral, the Chapter House, *c*.1275. In the top-plates – an edge-halved and tabled scarf, sometimes used with a transverse key. It is also used in Winchester Cathedral without the key, where it is also in the top-plates, supported by stone.

Fig. 237

Also from Salisbury Cathedral Chapter House. Chase-tenons with both shoulders spurred are used at all points where the arched braces meet the principal-rafters.

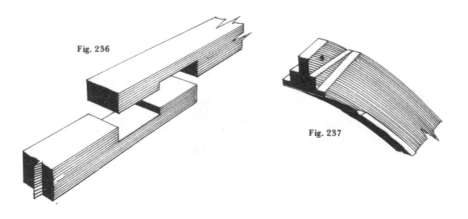

Fig. 236

Fig. 237

Fig. 238

From Place House, Ware, Hertfordshire. On the nomenclature the late Mr. S. E. Rigold stated that Place House is a proper name for a spacious secular residence intruded on ecclesiastical property – complaints about which name the Countess of Winchester, before 1235, and Joan de Bohun (d. 1283) – a date that must be considered when assessing this building. The scarf used for its top-plate may be defined as: stop-splayed and tabled with square under-squinted abutments, laterally keyed, with both terminal tables tongued and edge-pegged, with four face-pegs. The splay of this superlative joint measures 28½ ins. in length, and the joint is the finest of its kind yet known (Hewett 1980, 122-3).

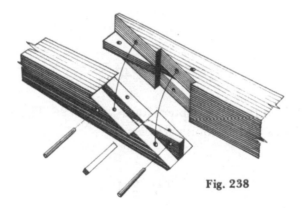

Fig. 238

Fig. 239a

From York Minster, the Chapter House, 1286-96. Shows a fished scarf trapped between two long timbers, with two face-pegs and a transverse long mortise.

Fig. 239b

Shows a splayed scarf with square under-squinted abutments, and three face-pegs, in the spire-mast.

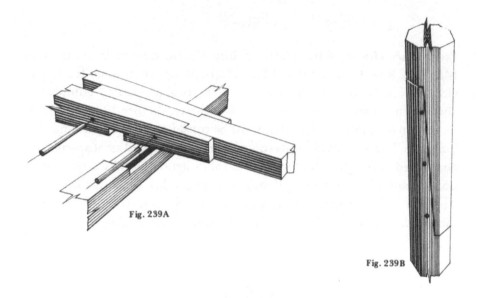

Fig. 239A

Fig. 239B

Fig. 240

From York Minster, the Chapter House, c.1286-96. The doors are made of seven planks with diminished fillets.

Fig. 241

From Wells Cathedral, the Chapter House. A through-splayed scarf, counter-tongued-and-grooved, with two edge-pegs. This joint was used for the sills of the roof, where it was subject only to compression. The work was completed by 1306, and this form of the scarf may therefore be considered as part of the exploitation of tongued-and-grooved splays that evidently preoccupied master carpenters during the reign of Edward I.

Fig. 242

From Wells Cathedral, the Choir, completed by 1310. A through-splayed and tabled scarf in the tie-beam, which was augmented by

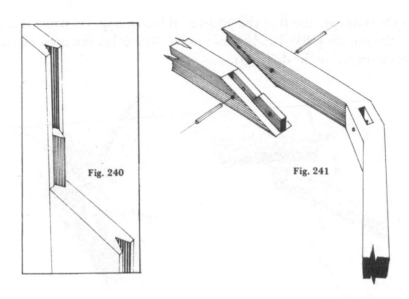

Fig. 240

Fig. 241

strap-irons forged into fleur-de-lys terminations and affixed by a fore-lock-bolt and five nails.

Fig. 243

From York Minster, the nave aisles, 1361-75. The sole-pieces of the common-rafters are trenched across the wall-stones.

Fig. 244

From Tewkesbury Abbey, Gloucestershire, the north transept. The Figures A and B are different in construction, A being at the top of the

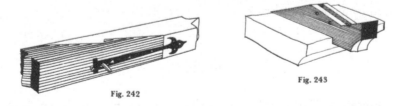

Fig. 242

Fig. 243

North Transept, and B at the bottom. B has an angular structure, so that the middle is fixed and cannot move. The other features are self-explanatory from the drawings.

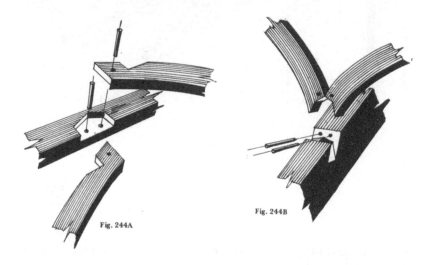

Fig. 244A
Fig. 244B

Fig. 245

From the Roof of No. 22, Vicars' Close, Wells, *c*.1363. All the purlins are fitted into the principal-rafters by means of tenons with soffit spurs, and in the case of the lowest ones this fitting was also scribed.

Fig. 246

From Carlisle Cathedral, the high-roof of the Choir, south side, by John Lewyn, 1363-95. The purlin scarfs are stop-splayed with square under-squinted abutments, one face peg and two iron spikes. The sole-pieces are trenched across their undersides to house a wall-plate, now replaced by a course of brick.

Fig. 247

From Hereford Cathedral, the south-east transept. The lower joint is the edge-halved scarf with bridled abutments, and various pegs. This

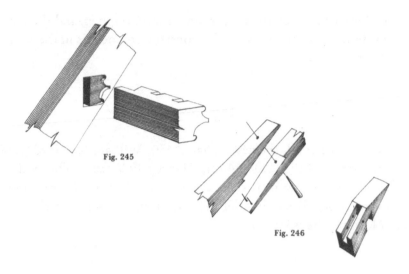

Fig. 245

Fig. 246

joint is first known in the sill of the London waterfront at Trig Lane, where it was dated archaeologically to *c*.1375. Such early examples are normally long in their halvings. The type continued in use until *c*.1650, the date of the last known specimen, in the barn at Rickling Green, Essex (Hewett 1969, 161-2).

Fig. 248

From Winchester Cathedral, the presbytery, 1315-60, by Thomas Witney. Mounted on the ties were four upright members, each of two

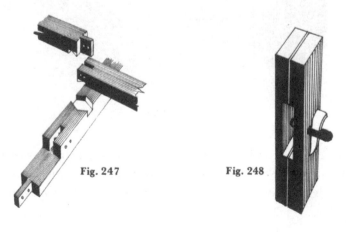

Fig. 247

Fig. 248

timbers, between which the scissor braces were trapped, and the whole forelock-bolted, as shown. A similar construction occurs in the roof of the north transept.

Fig. 249

Winchester Cathedral, the nave 1394-*c*.1450. William Wynford (1360-1403) or Robert Hulle (1400-1442), (Harvey 1972, 165). The evidence for these frames being intrusions consists in the fact that the straight braces were seriously cut in order that they might avoid the earlier 'soulaces'. Nothing else is known.

Fig. 250

From Wells Cathedral, the Vicars' Treasury. The floor joists are jointed with double tenons, the joists placed on edge, *c*.1420 (L. S. Colchester, verbal information).

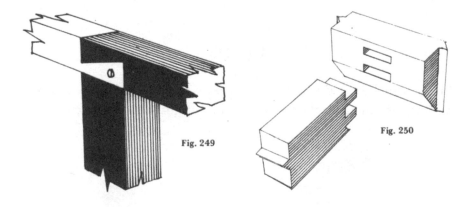

Fig. 249

Fig. 250

Fig. 251

From Wells Cathedral, the Library. The library roof was completed by 1433 (Hewett 1980, 195). Both the joints uniting the camber-beam and the ridge piece, shown at left of the figure, and that uniting the common rafters and the ridge-piece, are identical in principle; only the scribings are dissimilar. Definable as a central tenon of full width, with both face- and soffit shoulders forming one continuous spur bearing, this was evidently developed from the soffit spur, and must have been the point from which Richard Russell derived his diminished haunch.

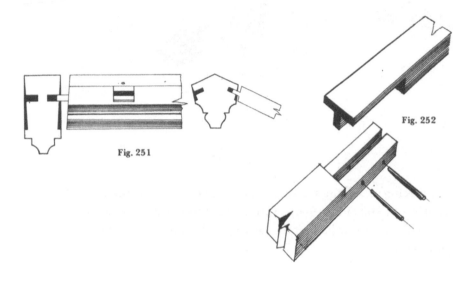

Fig. 252

Fig. 251

Fig. 252

Fig. 252

From Canterbury Cathedral, the north-west transept. The transept itself was designed by Richard Beke and built during the period 1448-55. The same scarf is used in the great barn at Harmondsworth, Middlesex, and that at Little Wymondley, Hertfordshire.

Fig. 253

From Norwich Cathedral, the north nave triforium. 'This certainly belongs to the raised triforium of the late 15th century, part of the masonry was in progress in 1472, so a date of *c*.1475 is the likeliest' (Dr. J. H. Harvey, personal communication).

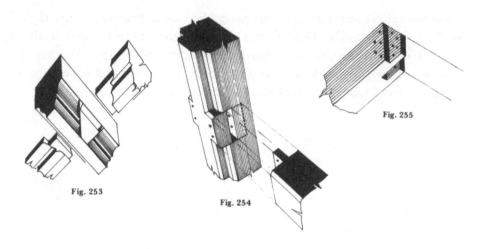

Fig. 253

Fig. 254

Fig. 255

Fig. 254

From King's College Chapel, Cambridge, the first phase, *c.*1480. The jointing of the side-purlins into the principal-rafters was effected by masons' mitres worked on the rafters, and with unrefined central tenons on the purlins.

Fig. 255

From King's College Chapel, Cambridge. Every rafter has a double mortise and tenon and 10 face-pegs.

Fig. 256

From King's College Chapel, Cambridge, the second phase, 1510-12. According to Dr. J. Saltmarsh, the framing of the remaining and greater part of this roof was resumed in March or April 1510, and completed in August 1512. During the interval the tenon with diminished haunch – a refined and much stronger joint – had been developed.

Fig. 257 and 258

From Lichfield Cathedral. In the general restoration by Sir William Wilson, 1661-9, a form of X lap-joint was used at the tie-beams, and the eaves triangles were secured by forelock-bolts with double washers.

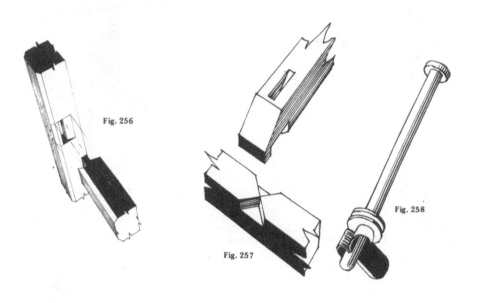

Fig. 256

Fig. 257

Fig. 258

Fig. 259

From Westminster Abbey, the south-east transept. A face-splayed scarf, counter-cogged and clamped by four edge-forelocks. The joint would not hold without the forelocks, there being no trapping. This has not been seen before, but the date is about *c*.1650-1700.

Fig. 260

From Winchester Cathedral, the east end of the nave, 1699. The king-post has strap-irons that are affixed with forelock-bolts.

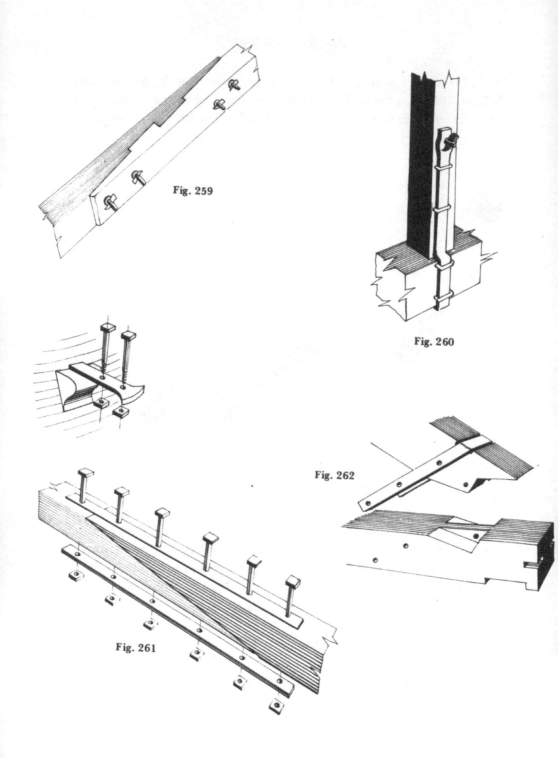

Fig. 259

Fig. 260

Fig. 262

Fig. 261

Fig. 261

From St Paul's Cathedral, 1675-1710. A through-splayed scarf with iron plates above and beneath, and six bolts with screw-threaded nuts.

Fig. 262

From St Paul's Cathedral. The joint at the feet of the principal-rafters, an unusual form of bridle-joint.

Fig. 263a and b

From Worcester Cathedral, the eastern arms.

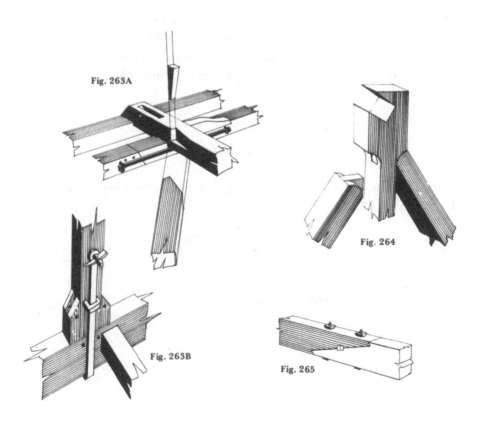

Fig. 263A

Fig. 264

Fig. 263B

Fig. 265

Figs. 264 and 265

From Lincoln Cathedral, the north nave triforium, 1762-65, by James Essex. The provision of 'compassed' butment-cheeks for the shoulders of the two struts is interesting, and apparently aims at two things, an entirely even and uniform distribution of the weight into the section of the struts, and also an abutment which would remain efficient if the post were to incline from the vertical. The scarfing is an attempt at progress; it derives from the mid-13th century splayed and tabled joints that were closely integrated by driving a wedge, but in this case the folding-wedges provided apparently sought to achieve the same effect without the tabling. It is twice bolted through.

Fig. 266

From Lincoln Cathedral, the Chapter House, 1761-2, by James Essex. Executed entirely in softwood. This structure relies upon ironwork and forelock-bolts, but cannot escape the basic unsuitability of pinewood for such complex carpentry; the timber was splitting badly at the time of inspection in 1972.

Fig. 267

From Rochester Cathedral, the great north transept, by L. N. Cottingham, 1825.

Figs. 268 and 269

From Sherborne Abbey, the choir roof by J. H. P. Gibb, 1856.

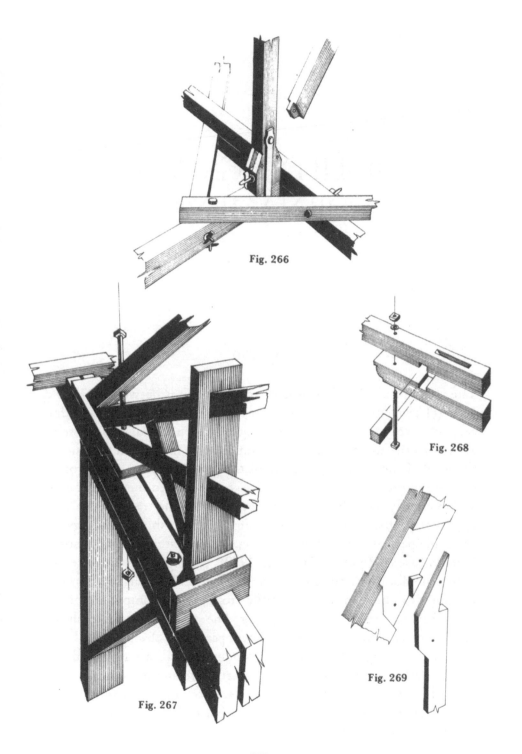

Fig. 266

Fig. 267

Fig. 268

Fig. 269

MOULDINGS AND ORNAMENT

Fig. 270

The transept of Winchester Cathedral was built between 1079 and 1093 (Harvey 1974, 243), and the second earliest piece of Cathedral carpentry yet recorded survives in the southern arm of this transept, affixed to the inside face of the south gable, immediately beneath the level of the existing tie-beam of the high-roof. It is a half-round timber column, or respond, that does not relate to the present roof and evidently dates from the first building of the transepts. It was, apparently, disguised with paint and painted ashlar-markings, to resemble part of the masonry – a deceit used later at parish church level in Essex (Hewett and Smith 1972). Owing to its great height above the floor it is difficult to examine, but from the adjacent tri-forium it can be seen that it is 'built' from several component pieces of timber, apparently nailed together. The capital is of the 'cushion' type, with a carinated fillet as the lower part of its abacus, and a bow-tell for the astragal. Comparable wooden capitals are recorded in a former prebendal hall in Essex, also ascribed to an early date (Hewett 1980, 45). It may be that this item related to the type of high-roof that provided a timber ceiling, such as that surviving above the nave of Peterborough Cathedral.

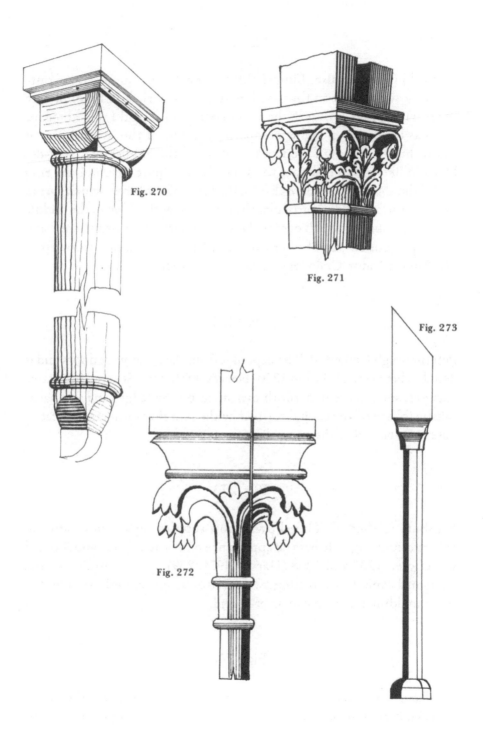

Fig. 270

Fig. 271

Fig. 273

Fig. 272

Fig. 271

The Old Court House, Limpsfield, Surrey. 'The capitals are of out-
standing sophistication for wooden domestic work, and I would date
them without hesitation to the years 1190-1200 ... and I would regard
them as the closest possible comparison to one of the round point at
Canterbury Cathedral, the work of William the Englishman 1178-86.
I would like you to compare the very classical 'pure' form of the ring
under the abacus, the square abacus, the form of the leaves which spread
over the surface, and in particular the leaf form with a blob in the middle
and a frill round it which occurs there – not very often elsewhere – and
is conspicuous on one of your capitals' (Personal communication from
Dr. Pamela Tudor-Craig, September 1972, to Mrs. Percy).

Fig. 272

Peterborough Cathedral. The capital is from the great west doors and is
dated to between *c.*1193 and 1230 (Harvey 1974, 235). Several decorative
features, such as the dog-tooth ornament on the ledges and the naturalistic foliage of the capital formed by the two shut-timbers, indicate a
date near the end of the given date range.

Fig. 273

Salisbury Cathedral. This shows a post supporting a valley-rafter in
the triforium angle. It has the appearance of a crown-post, and is dated
to between 1237 and 1258 (Harvey 1974, 239). The moulding of the
octagonal capital, comprising two cavettos, is surprisingly simple; the
base moulding is a simple quarter-round.

Fig. 274

St Albans Cathedral. The timber vault of the presbytery is the oldest
of its kind in England. The deeply convoluted moulding would fit into

the Early English period, except that the lower section using two cymas could easily be confused with late 15th- or early 16th-century work.

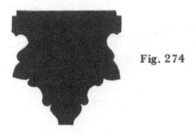

Fig. 274

Fig. 275

Bushmead Priory, Bedfordshire. The five crown-posts of the roof, shown with the carpenters' assembly marks scribed on them, are all different. The only common feature is the form of the astragals. It stands as a lesson that caution is always necessary when dating even with well-developed mouldings.

Fig. 276

Salisbury Cathedral. The Sacristy, now called the Vestry, 1260-80, restored *c*.1863 (verbal information from Mr. R. Spring, Clerk of Works). The consistent feature of the capital is the use of the half-round with frontal fillet. The base has a crude water-holding groove.

Fig. 277

Salisbury Cathedral. The Chapter House, *c*.1275 (Harvey 1974, 239). The capital bears the undercutting common in the 13th century. However, the cavetto and roll is beginning to appear, a moulding more at home in Decorated and Perpendicular contexts. The late 13th century is appropriate here.

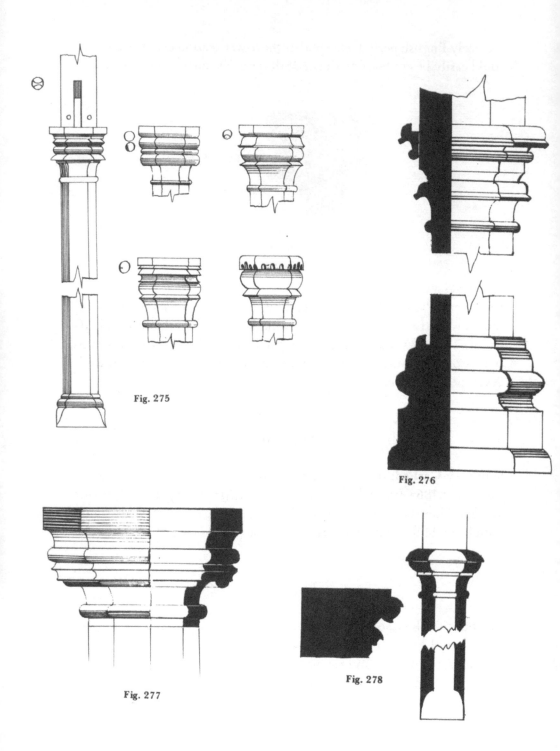

Fig. 275

Fig. 276

Fig. 277

Fig. 278

Fig. 278

Bisham Abbey, Berkshire, the crown-post of the hall. The only moulding here is the scroll common to the late 13th and early 14th centuries.

Fig. 279

Prittlewell Priory, Essex, the Prior's Chamber. This illustrates three
features. At *a* is shown the top and bottom of one of the crown-
posts, the base being kept square, as at Priory Place, Little Dunmow,
Essex (Hewett 1980, 129-30). *b* shows the base of one principal-post;
and at *c* a section of tie-beam is shown. The base moulding of the
crown-post, raised a little above the tie-beam, is paralleled by other
examples of approximately the same period, such as Place House, Ware,
Hertfordshire, *c.*1295 (Hewett 1980, 122-3). The characteristic bell base
of later medieval times is emerging.

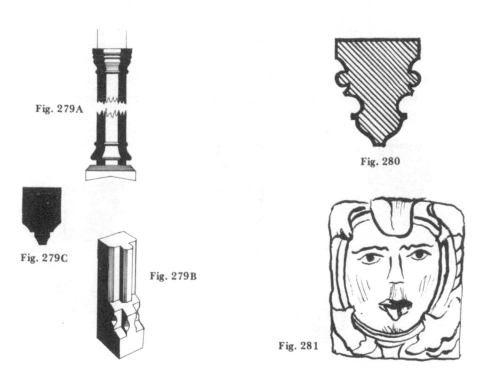

Fig. 279A

Fig. 279C

Fig. 279B

Fig. 280

Fig. 281

Figs. 280 and 281

Rochester Cathedral. The South Choir aisle, which in St John Hope's view, came into its present state shortly after 1322; a section of one of the rib timbers, not drawn to scale. The use of a cavetto enclosing a roll becomes very common later; the depth of the moulding helps to place this quite early. This face is one of a group cut on the central bosses. They are excellently done, which is common to the emerging Perpendicular style.

Fig. 282 and 283

Ely Cathedral, the Octagon. This shows a moulding exactly dated between 1328 and 1342. It should be compared with that in Fig. 280, with which it is remarkably similar. The wall-plate moulding with a scroll is a warning not to date too late.

Fig. 284

From Cartmel Priory, Lancashire. This is the hollow chamfer which Forrester ascribes to *c.*1370-1550, but here I believe it is used earlier, *c.*1340 (Forrester 1972, 31).

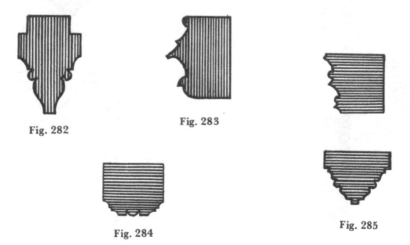

Fig. 282

Fig. 283

Fig. 284

Fig. 285

Fig. 285

Gloucester Cathedral. From the Cloister, *c.*1370-1412. The wall-plate moulding has two rolls in cavettos, showing the ultimate regular use of this moulding. The lower moulding, of the door muntin, shows the double ogee here split by a roll – the double ogee comes into high fashion in the 15th century.

Fig. 286

Winchester College, from the doors called Middle Gate, dated to 1393. These mouldings are somewhat unusual in having a central quirk in a cavetto. The lower part of the transom moulding is more unusual in having a roll in a cavetto.

Fig. 287

Wells Cathedral, Bishop Bubworth's Library, 1433.
(a) The hollow chamfer, *c.*1370-1550, when doubled in stone. The earliest part of the moulding is the profile at the base (Forrester 1972, 34, no. 74, Worcester Cathedral, nave and North Arcade, *c.*1320).
(b) Early in the 14th century new base profiles are introduced. Sometimes both of the upper rolls are abandoned for an ogee-like moulding (Forrester 1972, 36, no. 117, Westminster Abbey, detail from west bay of nave arcades, late 14th century).

Fig. 286

Fig. 287

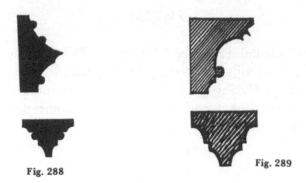

Fig. 288

Fig. 289

Fig. 288

Durham Cathedral. From the door in the Galilee, 1428-35. The use of more than one roll separated in a cavetto is expected in the early 16th century. However it can and does occur in the earlier 15th century.

Fig. 289

Norwich Cathedral. From the roof of the north nave triforium, which no longer exists. These two mouldings fit well into the later 15th century – the very deep cavetto and roll in the upper section, and the cavetto, return and cavetto in the lower drawing are both typical.

Fig. 290

Winchester, the Hospital of St Cross. The use of multiple deep cavettos puts this in the Perpendicular period. Its lavishness suggests a late appearance in this period.

Fig. 291

Abbey Dore, Herefordshire, from the roof by John Abel, 1633-4. The pendant having a post-medieval shape is cut like a cage, typical of early 17th-century work.

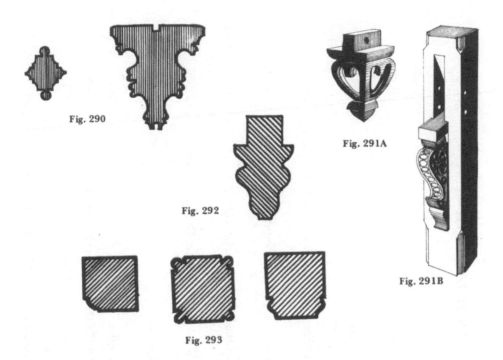

Fig. 290

Fig. 291A

Fig. 292

Fig. 291B

Fig. 293

Fig. 292

York Minster, the south main transept by L. Tebb, *c.*1770-80. The slackness of the profile has left the medieval vigour far behind. This is a poor moulding.

Fig. 293

St Paul's Cathedral (1675-1710). These three mouldings really only produce shadow lines in the corners. They have very little impact.

Note: These are observations on drawings of mouldings shown to me by Cecil Hewett at a time when he was unable to express himself easily. In my role as an amanuensis I am reasonably sure that these are his thoughts as well as my own.

Adrian Gibson

DEVELOPMENT OF NOTCHED LAP JOINTS

Building	Joints	Architect	Date
Rayleigh Castle, Rayleigh, Essex	Stub-mortise	—	1066-1086
'Norman Hall', Peterborough Cathedral	OPEN ARCHAIC notched lap	—	c.1155-77
Waltham Abbey, Waltham, Essex	Archaic notched laps, all the same	—	c.1180
Coggeshall Abbey, Essex, barn	REFINED open notched laps	—	(1130±60) c.1160
Canterbury Cathedral, south-east spirelet	SECRET notched lap but archaic profile	William of Sens (1174-d.1180)	c.1184
Lincoln Cathedral, St Hugh's Choir	Better long archaic profile, notched laps	(Geoffrey de Noyers) Richard Mason (fl. c.1195)	1195-1200
Wells Cathedral, west part of nave	SECRET and REFINED notched laps	Adam Lock (d.1229)	1209-1213, during the Interdict
Salisbury Cathedral, north-east transept	Secret but not refined profiles	(Elias of Dereham), Nicholas of Ely	c.1225-37
Romsey Abbey, Hampshire, refectory	Secret archaic with refined profiles	—	c.1230
Whepstead barn, Suffolk	Using notched laps and a mortise	—	c.1255
Westminster Abbey, eastern apse	Refined profile and DIMINUTION	Master Alexander	1245-59
Exeter Cathedral, the transepts	Between two notched laps is a counter-sallied cross-halving	Thomas Witney	c.1325
Wells Cathedral, south choir transept	Refined notched laps, with diminution by its edge	William Joy (1329-46)	1330

CROWN-POST WITHOUT COLLAR-PURLIN

Building	Architect	Date
Chichester Cathedral, nave, west end	?Walter of Coventry	1187-99, Dr. J.H.H.
Salisbury Cathedral, north porch	(Elias of Dereham) Nicholas of Ely	porch, c.1250-1237-1258, Dr. J.H.H.
Winchester Castle, roof	Stephen	1222-1236, Dr. J.H.H.
Westminster Abbey, north transept	Alexander	1245-1259 Colvin 1963
Westminster Abbey, choir	Alexander	1245-59
Shere Church, Surrey, west transept	–	1220-36, C.A.H.
Bushmead Priory, Beds., nave	–	c.1280, Bony, 1979
Bisham Abbey, Berks., hall	–	c.1280 Fletcher and Hewett 1969
Winchester Cathedral, presbytery and north transept	Thomas of Witney	c.1315-1360, Dr. J.H.H.
Beeleigh Abbey, Essex	–	c.1400, C.A.H.

WITHOUT TIE-BEAMS AND WITHOUT SOULACES

Building	Architect	Date
'Norman Hall', Peterborough Cathedral precinct	–	c.1155-1177, C.A.H.
14, St Paul's Street, Stamford, Lincs.	–	–
The Chapel, Harlowbury, Harlow, Essex	–	–
St Mary Magdalene, East Ham Church, Essex	–	–
'Middle Saxon' "Palaces", Northampton	–	–
Chappel Church, Essex	–	–
St Mary the Virgin Church, Lindsell, Essex	–	–
West Range, Ely Cathedral	–	–

WITHOUT TIE-BEAMS, WITH SOULACES

St Martin's Church, Canterbury	–	–
St Albright's Church, Stanway, Essex	–	c.1125-50, R.C.H.M.
St Martin of Tours Church, Chipping Ongar, Essex	–	c.1125-50, R.C.H.M.
Peterborough Cathedral (re-used)	–	c.1155-75, Dr. J.H.H.
Greyfriars, Lincoln nave, eastern part	–	c.1237, Martin, 1935
Coggeshall Abbey, Essex, Capella-extra-portas	–	c.1218-1223, Gardner, 1955

Building	Architect	Date
Greyfriars, Lincoln, west end	–	c.1260
Blackfriars' Church, Gloucester	–	c.1267, Rackham et al., 1978
Exeter Cathedral, presbytery	William Luve	1308-17, Dr. J.H.H., 1972
Ottery St Mary's Church, Devon, nave	–	c.1342, C.A.H.
Carlisle Cathedral, choir	?John Lewyn (1364-d.c.1398)	1363-95, Dr. J.H.H.
Hereford Cathedral, Lady Chapel	–	1217-25, Dr. J.H.H.
York Cathedral, south transept	–	1770-80

TIE-BEAM TO EVERY RAFTER COUPLE

Waltham Abbey, Essex	–	c.1150, C.A.H.
Little Hormead Church, Herts	–	c.1150, C.A.H.

TWO RAFTER COUPLES BETWEEN THE TIE-BEAMS

Wells Cathedral, east nave	?Adam Lock, d. 1229	1175-80
Wells Cathedral, west nave	?Adam Lock, d. 1229	after 1213
Lincoln Cathedral, St Hugh's Choir	Richard Mason (fl. c.1195)	1195-1200, Dr. J.H.H.

Lincoln Cathedral, chapter house vestibule	Alexander	c.1220-35, Dr. J.H.H.
Lincoln Cathedral, Morning Chapel	Alexander	1220-30 Dr. J.H.H.
Lincoln Cathedral, north choir transept	Alexander	c.1200, Cook, 1970
Lincoln Cathedral, nave	Alexander	c.1225-53, Dr. J.H.H.
Lincoln Cathedral, Angel Choir	Simon of Thirsk (c.1260-90)	1256-80 Dr. J.H.H.

THREE RAFTER COUPLES BETWEEN THE TIE-BEAMS

Lincoln Cathedral, south choir transept	(Geoffrey de Noyers) Richard Mason (fl. c.1195)	c.1200-1210
Salisbury Cathedral, north-east transept	(Elias of Dereham) Nicholas of Ely	c.1225-37, Dr. J.H.H.
Winchester Cathedral, nave, eastern part	–	c.1300, C.A.H.
Winchester Cathedral, south transept	?William Lyngwode	?c.1300-10
Exeter Cathedral, nave	Thomas Witney	1328-42, Dr. J.H.H.

FOUR RAFTER COUPLES BETWEEN THE TIE-BEAMS

Building	Architect	Date
Bushmead Priory, Beds	–	c.1280, Bony 1979
Peterborough Cathedral, Table Hall	–	c.1450, C.A.H.
Beverley Minster, great south transept	?N. Hawksmoor	1715-36, C.A.H.

FIVE RAFTER COUPLES BETWEEN THE TIE-BEAMS

Beverley Minster, nave, part	–	c.1213-60
Boxgrove Priory, Sussex	–	early 12th century, R. Ratcliff

Wells Cathedral, 'Chequer'	?Thomas Witney	c.1320, Mr. J.S.C.
Bisham Abbey, Berks (east wing)	–	c.1336, Dr. J.M.F.
Bristol Cathedral, choir	?William le Geometer	c.1340, Dr. J.H.H.
Wells Cathedral, No. 22 Vicars' Close	–	c.1360, J.S.C.
Oxford Cathedral, Latin Chapel	–	c.1355, Dr. J.H.H.
Durham Cathedral, monks' dormitory	John Lewyn, Thomas Mapilton, (1408-d.1438)	1398-1404, C.A.H.
Canterbury Cathedral, north-west transept	Richard Beke (1409-d.1458)	1448-55, Dr. J.H.H.
Canterbury Cathedral, south-east transept	–	1771, dated
Salisbury Cathedral, nave	James Wyatt (1746-1813)	1787-93, Dr. J.H.H.
Ely Cathedral, east front restored	James Essex (1722-d.1874)	1757-62, Dr. J.H.H.
Durham Cathedral, choir	James Clement	1690, Dean of Durham
Canterbury Cathedral, south-west transept	George Austin	1834-41

SIX RAFTER COUPLES BETWEEN THE TIE-BEAMS

Westminster Abbey, north transept	Alexander	1245-59, Colvin 1963
Winchester Cathedral, north transept	Thomas Witney	1315-56, Dr. J.H.H., C.A.H.
Worcester Cathedral, eastern arms	–	c.1700, C.A.H.
Lichfield Cathedral, south transept, Lady Chapel	Sir William Wilson (1641-d.1710)	1661-69, Dr. J.H.H.

SEVEN RAFTER COUPLES BETWEEN THE TIE-BEAMS

Salisbury Cathedral, north porch	(Elias of Dereham), Nicholas of Ely	c.1250, Dr. J.H.H.
Chichester Cathedral, nave	?Walter of Coventry	c.1187-89, Dr. J.H.H.

Tewkesbury Abbey, north transept	–	–

Building	Architect	Date
Winchester Cathedral, presbytery	Thomas Witney	c.1315-60, Dr. J.H.H.
King's College Chapel, Cambridge, nave, western end	Richard Russell	1510-12, Dr. J. Saltmarsh
Winchester Cathedral, nave, western end	–	1699 (W.J. C-Turner)

EIGHT RAFTER COUPLES BETWEEN THE TIE-BEAMS

Wells Cathedral, south choir transept	?Thomas Witney (1316-56)	before 1330, Dr. J.H.H.
Worcester Cathedral, nave	John Clyve (1362-92)	c.1360-74, Dr. J.H.H
Bath Abbey, nave	Robert Vertue (1475-d.1506) William Vertue (1500-d.1527)	1501-39, Dr. J.H.H.
Oxford Cathedral, nave	–	1816, J.B.

NINE RAFTER COUPLES BETWEEN THE TIE-BEAMS

Lincoln Cathedral, Consistory Court	Alexander	1225-35, Dr. J.H.H.
Gloucester Cathedral, Lady Chapel	?John Hobbs	1457-83, Dr. J.H.H.
York Minster, choir, nave west	Sir Robert Smirke, Sydney Smirke	1829-32, 1840
Salisbury Cathedral, Old Deanery	–	1258-74, Drinkwater 1964
Bisham Abbey, Berks., hall	–	c.1280 Fletcher and Hewett, 1969

WITH MULTIPLE PURLINS

Denny Abbey, nave	–	c.1230, C.A.H.
Tewkesbury Abbey, nave	–	–

Beeleigh Abbey, Essex, dormitory	–	c.1400, C.A.H.
Wells Cathedral, north transept	–	1661, L.S. Colchester
Abbey Dore, nave	John Abel	1633-4, Colvin 1948
St Paul's Cathedral, nave	Sir Christopher Wren	1696-1706, Lang 1956

QUEEN-POSTS WITHOUT PURLINS

Lincoln Cathedral, St Hugh's choir, east transept	(Geoffrey de Noyers) Richard Mason (fl. c.1195)	1192-1200
Salisbury Cathedral, choir transept	(Elias of Dereham) Nicholas of Ely	1225-1237

Building	Architect	Date
Tewkesbury Abbey, north transept	–	c.1260
Winchester Cathedral, nave		c.1300
Abbey Dore, Heref., nave	John Abel	1633-4, Colvin, 1948
Wells Cathedral, north transept	–	1661
Lincoln Cathedral, north-west transept	James Essex	1762-65
Ely Cathedral, choir	James Essex (1722-d.1784)	1768
Lincoln Cathedral, great transepts	Restoration	?1762-65
York Minster, nave	Sydney Smirke	1840
Sherborne Abbey, west choir	–	?1849
Sherborne Abbey, east choir	–	1856

KING-POSTS

Westminster Abbey, the apse	Alexander	1245-59, Colvin, 1963
Salisbury Cathedral, Old Deanery	–	1258-74, Drinkwater 1964

Worcester Cathedral, nave	John Clyve (1362-92)	1375-95
Durham Cathedral, monks' dormitory	John Lewyn; Thomas Mapilton (1408-d.1438)	1398-1404
Hereford Cathedral, Vicars' Cloister	–	c.1472
Lichfield Cathedral, east transept	Sir William Wilson	1661-69
Winchester Cathedral, west nave	–	1699
Tewkesbury Abbey	–	–
Worcester Cathedral, nave	–	late 14th century, C.A.H.
Worcester Cathedral, eastern arms	–	c.1700, C.A.H.
Beverley Minster, south transept	N. Hawksmoor	1715-40
York Minster, south transept	L. Tebb	1770-80
Canterbury Cathedral, nave	–	1771
Rochester Cathedral	L. N. Cottingham	c.1825
Ripon Cathedral, nave	'C.S.R.'	1904

GLOSSARY

Addorsed: From heraldry, meaning placed back to back–the opposite of 'addressed', or face to face.

Angle tie: Tying timbers placed across angles, normally the returns of wall plates. These were widely used during the 18th century as a means to step hip rafters, which were seated in a third timber, the dragon piece.

Abutment: *Abut*, O.Fr. *abuter*, to touch at the end (*à*, to, *bout*, end). Any point in timber jointing where one timber's end touches another constitutes an abutment. A 'butt-joint' is, therefore, one where ends meet; no integration is implied.

Anglo–Norman Romanesque: A period of English architectural history during which the style was based on Roman buildings having round, or single centred arches, covering the period from the Conquest in 1066 until *c*.1200.

Anglo–Saxon Romanesque: The period of English architectural history covering the years between *c*.1000 and the Norman Conquest in 1066.

Arcature: The curvature of an arch, as segmental, ogee or lancet.

Arcade: A range of arches. Term applied also to the series of posts standing inside an aisled timber building, because they are sometimes arch-braced in their longitudinal direction.

Arch braces: Term generally applied to braces beneath tie beams, which were frequently curved, or arched.

Arris: The edge at which two surfaces meet.

Arris-trenched: Trenched (q.v.), so that the trench is cut obliquely through an arris and affects both adjacent surfaces.

Ashlar pieces: Short, vertical timbers at the feet of rafters, generally standing upon sole pieces. These continue the internal wall surface until it meets the underside of the rafters, avoiding a visual discontinuity and greatly strengthening the rafters' base.

Barefaced: With the face uncovered, without a mask; avowed, open. Term used to denote a timber joint possessed of only one shoulder, but which normally possesses two.

Base crucks: Timbers placed as wall posts and containing the naturally grown angle of the eaves, above which they may rise to collar height.

Bays: The divisions, normally postulated by the material used for construction of the lengths of buildings. In the case of arcades, each arch is taken as one bay.

Bird's-mouth: Term used to describe joints bearing a visual resemblance to an open bird's beak.

Blade, -ing, -ed: Term used to specify scarfs that are face-halved and terminated in inset, barefaced tongues.

Bole: The butt of a tree trunk, normally of concave conoid form, used to provide jowls by inverting the timber.

Bowtell: Small roll moulding, or bead.

Brace: Any timber reinforcing an angle, usually subjected to compression.

Bridle: Term applied to timber joints having open-ended mortises and tenons resembling a horse's mouth with the bit of the bridle in place.

Bridle-butted scarf: That category of scarf, or end-to-end joints for timbers which was in use throughout the Perpendicular period. An example is shown in Fig. 23.

Bressummer: Breast-summer, a timber extending for the length of a timber building, normally forming the sill of a jettied storey.

Bridging-joist: Floor timber that supports the ends of common joists, and normally bridges the bays from one binding-joist to the next.

Broach: A spit or point.

Butment cheeks: The timber left on either side of mortises, against which the shoulders of tenons abut.

Butts: The ends of jointed timbers, or those parts of the edges of timbers touched by ends, and constituting abutments.

Butt-joints: The category of timber joints in which neither piece penetrates the other. When assembled the components merely touch, without any integration. They are held in contact by other timbers, or irons.

Butt-notches. Type of jointing for timbers which is shown at 'a' in Fig. 3. The notch is the nick, or indentation formed by the two converging cuts in the face of one timber, and the butt is the suitably shaped end of the other timber. These joints can only resist compression, and are archaic.

Camber beams: Beams sufficiently cambered to form the basis of the simplest type of roof, their curvature serving to drain their surfaces when clad.

Cant: The oblique line or surface which cuts off the corner of a square or cube. The term is applied to soulaces in roofs, because they produce a canted plane;

roofs possessing soulaces, collars and ashlar pieces are thus described as 'of
  seven cants'.
Cant post: Posts that converge upwards; see Navestock belfry.
Chamfer: The slope or bevel created by removing a timber's corner or arris.
  These are termed 'through' if run off the end of the workpiece, 'stopped' if
  terminating in a decorative form before the end of the piece or its conjunction
  with another.
Chase: From *chasse*, a shrine for relics. In carpentry a score cut length-wise, a
  lengthened hollow, groove, or furrow.
Chase, or chased mortise: A long mortise into which a tenon may be inserted
  sidewise.
Chase-tenon: a tenon which can be inserted into its (chase-) mortise in two ways,
  lengthwise or laterally. Two such tenons are shown in Fig. 7 on the sole- and
  ashlar-piece.
Cladding: The external covering applied to a wall or a roof.
Clamp: A term applied variously to timbers depending upon the type of building.
  In houses the term denotes horizontal timbers attached to the wall studs in
  order to support floors; these clamps normally indicate the later intrusion of
  such floors.
Clench, clinch: Either to turn the point of a nail or spike, and re-drive it back
  into the timber through which it has passed; or to form its end into a rivet, or
  clench, by beating it out upon a washer or rove.
Coak: A peg or dowel of a diameter almost equal to its length, used in
  19th-century shipwrighting to join futtocks and timbers because it was cheaper
  than scarfing them.
Cogging: A method of housing an entire timber's end, sometimes used to prevent
  its rotation—as in door cases.
Collar: A horizontal member placed across a rafter-couple, between their base
  and their apex, and considered to be in compression, normally. An example is
  shown at lower right in Fig. 5.
Collar beam: A roof timber, placed horizontally and uniting a rafter couple at
  a point between the bases and the apex. Collar beams can act either as ties
  or strainers.
Collar-purlin: A lengthwise timber in a roof assembly, of which an example
  is shown in Fig. 9. It is carried by vertical crown-posts, and connects the
  successive rafter-couples, longitudinally, by their collars. Its purpose was the
  lengthwise stability of the whole roof.
Compass timber: A term denoting timber of natural and grown curvature, as
  distinct from relatively straight-grown timber from which curves are cut.
Common joist, -rafter: The majority of either kind, and normally those of the
  least cross section in any floor or roof.
Corner post: The post standing at the return of two walls, as at the end and
  adjacent side of a building.

Counter rebate: See Fig. 57.

Crenellate: To furnish with battlements, a decorative device used much in timber buildings.

Crown-post: Vertical timber posts, much carved and decorated in Perpendicular times (see Fig. 15), and plain functional posts at their inception, as shown in Fig. 9. These stood, normally, upon tie-beams and carried the collar-purlin.

Cruck blade: The elbowed timber forming one half of a pair of crucks.

Cusp: In Gothic tracery the pointed shape or form created by the intersection of two concave arcs.

Cyma: Ogee. Formed by a concave and a convex arc in a single linear association.

Decorated, period: That period of English architectural history covering the years between *c*.1250/1350.

Dormer: An upright window protruding from the pitch of a roof.

Double tenons: Two tenons cut from the same timber's end and placed in line; if side by side they constitute a pair of single tenons.

Dragon beam, -piece: A timber bisecting the angle formed by two wall plates. If a beam it supports a jetty continued around the angle; if a piece it normally serves to step a hip rafter.

Draw knife, or -shave: A hand tool possessing handles at either end of its blade, used to produce chamfers.

Durns: Timbers with grown bends suitable for the manufacture of door-ways of Gothic arcatures; two were frequently sawn from one piece of the requisite form.

Early English, period: That part of our architectural history approximating to *c*.1150-1250.

Eaves: The underside of a roof's pitch that projects outside a wall.

End girt, -girth: Horizontal timber in an end wall placed halfway betwixt top plate and groundsill, thereby shortening the studs and stiffening the wall.

Fascia: A board forming a front, frequently used to cover a number of timbers' ends, as joists at a jetty, or rafters at an eaves.

Fillet: In moulding profiles a small raised band, normally of square section. Also a small squared timber.

Fish: A length of timber with tapering ends which can be used to cover and strengthen a break in another timber.

Fished scarf: A scarf that relies upon the introduction of a third timber.

Footing: Foundation.

Free tenon: A tenon used as a separate item, both ends being fitted into mortises cut into two timbers to be joined; often used to effect edge-to-edge joints.

Gambrel: Perhaps Old French (Norman). 'Also gambrel roof . . . so called from its resemblance to the shape of a horse's hind leg' (*O.E.D.*, 1933).

Girth, girt: Horizontal timbers in wall frames, placed at half height, which shorten and thereby stiffen the studs.

Groundsill: The first horizontal timber laid for a timber building. As the name implies these were in ancient times laid directly upon the ground, as was the case at Greensted church, Essex.

Halving: In jointing the removal of half the thicknesses of two timbers, as in cross halving.

Harr, arr: The edge timber of a door leaf or gate nearest the hinges, the opposite edge timber of which is the head.

Hanging knee: Term denoting a knee placed beneath a beam. Knees placed above beams are 'standing knees, or standards', and those in the horizontal plane are 'lodging knees'. All three derive from shipwrighting and were used in ships as early as the Viking period.

Haunch: Adjuncts of tenons, designed to resist winding; they may be square or diminished.

Header: Short timber to carry rafters' tops at the exiture of chimney-stacks.

Hewn knee: A knee, or angle, cut from a timber's end, as distinct from a separate and applied piece.

Hip rafter: A rafter pitched on the line of intersection of two inclined planes of roof, forming the arris of a pyramidal form.

Hogging: Stress caused by supporting the centre of a beam and leaving the ends unsupported, as when a wave rises amidships of a vessel, and beneath her.

Housing: In jointing a cavity large enough to hold an entire timber's end.

Jacobean: The period between 1603 and 1625–the reign of James I.

Jamb: The side of a doorway, archway or window.

Jetty: The projection of a floor outside its substructure, upon which the next storey was built. This resulted in floor areas that increased as the storeys ascended.

Joggle: Said of two timbers, both of which enter a third but which have their joints out of line in order to avoid excessive weakening.

Joist: The horizontal timbers supporting floors; these are binding, bridging, common and trimming. Binders unite storey posts; bridgers span or bridge each bay from binder to binder; commons are the most numerous and actually carry the floor boards. Trimmers are used to frame the edges of voids, such as stair wells.

Jowl: (Jole) 'The external throat or neck when fat or prominent... the dewlap of cattle'. (O.E.D.) Term applied to the thickened ends of such timbers as storey posts which facilitate the jointing of several other timbers.

Kerf: The cut produced by a saw.

Key: Tapered piece of dense hardwood transfixing a scarf, used to close its abutments.

King stud: A stud placed centrally in a gable, normally supporting the collar purlin.

Lap dovetail: That form of dovetail that overlaps, and is not finished flush. The alternative form is the 'through dovetail' used by cabinet makers; they may also be 'secret', 'secret-mitred' or double.

Lap joints: Any jointed timbers which overlap each other.

Lodged: 'To put and cause to remain in a specified place.' (*O.E.D.*) A term applied to floors retained in place by their weight alone.

Main span: In aisled buildings this is the central and greatest distance spanned.

Midstrey: The porch-like structure at the front of a barn, derived from middle-strey. Each bay of a barn was a strey, and ancient barns normally had one such porch at their centre.

Mitre: Abutments at 45 degrees, producing square returns.

Muntins: Vertical members of panelled areas; the term may derive from mountants.

Mullions: Vertical components of windows, placed in the void.

Nogging: The material used to infill a framed wall betwixt sill and top.

Notched laps: A category of lap joints having V-shaped indentations on plan-view to prevent their lengthwise withdrawal.

Outshot, outshut: An area of space added to a building's bays, normally at the sides: when at the end of a building they are called by the ancient term 'culatia'.

Passing brace: A brace uniting several successive members of a frame and passing them by means of halved jointing; of mainly Early English and Decorated usage. (Author's coinage, 1962, without historical validity.)

Perpendicular period: That part of English architectural history covering, approximately, 1350-1450.

Plate: A horizontal timber laid at the base of a timber frame; the term implies a footing, as distinct from a groundsill.

Prebend: The share of the revenues of a cathedral or collegiate church allowed to a clergyman who officiates in it at stated times.

Prick post: Any vertical timber placed in compression, but not a storey post.

Principal-rafter: A heavy rafter placed at bay intervals, normally associated with side purlins.

Purlin: A longitudinal timber in a roof.

Queen posts: Posts set in pairs between tie beams and collars and acting in compression.

Raking struts: Inclined struts used in pairs between tie beams and principal-rafters.

Reversed-assembly: Indicates a system of rearing transverse framing units, the lengthwise timbers of which (top plates) are laid last. In these cases the tie beams are *under* the top plates. (Author's coinage, 1962, without historical validity.)

Rive: To split timber lengthwise, i.e. along its grain.

Rove: The circular plate, or washer, upon which the clench, or rivet, in boat- or ship-building is formed.

Sagging: Stress caused by supporting the ends of a timber and applying weight to its centre.

Sally: An obtusely angular and pointed projection, normally on a timber's end. Alternatively a 'tace'.

Samson post: 'Pillar erected in a ship's hold, between the lower deck and the Kelson'. (*O.E.D.*) The term alludes to the strength of Samson (Judges XVI, 29), and is applied to similar posts used to support early floors.

Scarfing: The jointing of relatively short timbers into continuous lengths, by means of various expedients; the four faces are smooth and continuous.

Scarfed cruck: A cruck blade having a scarf-jointed angle, as distinct from a grown angle.

Scissor-braces: Timbers crossed in a saltire and connecting a pair of rafters, illustrated in Fig. 3, at 'b'.

Secret notched-lap: A development of the notched lap-joint, in which the notch itself cannot be seen, since covered by a flange of timber. An example of the socket cut to receive one of these is shown in Fig. 7, where it is the third joint from the base of the rafter shown at the right.

Set: The divergence of the sides of a dovetail.

Seven-cants: The seven short, straight sides of the partial polygon built into certain timber roofs. Both 'a' and 'b' examples shown in Fig. 3 have seven cants. Such frames have been called 'trussed-rafters' during the last century.

Shore: An inclined timber supporting a vertical one, acting in compression.

Shuts: The edges of a door leaf collectively form the 'shuts' of that door.

Side girt, girth: See end girt.

Soffit: Underside or archivolt.

Sole pieces: Short horizontal timbers forming the base of any raftering system that has a base triangulation.

Soulace: A definitive term (Salzman, 1952) for secondary timbers connecting rafters with collars, and placed *under* the latter.

Splayed-scarf: A scarf, or end-to-end timber joint, which is effected by diminishing both ends to either nothing, or to a butt. An example is shown at lower left in Fig. 8.

Spur tie: A short tie such as connects a cruck blade and a wall plate, or a collar arch and a wall plate.

Spire mast: Central vertical timber of a framed spire.

Squint: Angle other than 90 degrees.

Storey post: A wall post of a multi-storeyed timber building that continues through the floor levels.

Straining beam: A horizontal beam between two posts, acting in compression to keep them apart.

Strut: A timber in a roof system that acts in compression, in a secondary capacity.

Stub tenon: A short tenon that does not entirely penetrate the mortised concomitant timber.

Studs: From O.E. *studu*, a post; the vertical common timbers of framed timber walls.

Table: A raised rectangular portion on a worked timber, normally a scarf adjunct.

Tace: See sally.

Tail: The male part of a dovetail joint.
Tie beams: Beams laid across buildings to tie both walls together; they must have unwithdrawable end joints for this purpose.
Tongue: A fillet worked along the edge of a plank to enter a groove in another.
Top plate: A horizontal timber along the top of a framed wall.
Trait-de-Jupiter: The early form of scarf-joint which is shown in Fig. 9; so named in medieval times because of its visual resemblance to lightning, or the zig-zag.
Transom: A cross beam acting as a support for the superstructure.
Trench: A square sectioned groove cut across the grain.
Trussed, Truss: In roof-framing trusses are the most rigid and stable frames, such as those having tie-beams in Fig. 9. Main-frames would be a more defensible term, however.
Tusk: The wooden key driven through the protruding end of a tusked tenon, an unwithdrawable form of that joint.
Waney: Used to describe timber, the squared section of which is the greatest that can be cut from the rounded trunk, when any missing sharp arrises are said to be 'waney' edges.
Winding: In carpentry the result of torque or twisting, or the result of drying a spirally-grained tree.
Wind braces: Braces fitted into the angles of either roofs or walls to resist wind pressures.

Only terms that are likely to be strange to the general reader are here given, and the correct reference is, of course, the *Oxford English Dictionary*.

INDEX

I Persons and Places

II Subjects and Terms